THE USE OF HANDBOOK
TABLES AND FORMULAS

This book should be used
in conjunction with the
nineteenth edition of
MACHINERY'S HANDBOOK

THE USE OF HANDBOOK TABLES AND FORMULAS

FIVE HUNDRED EXAMPLES AND TEST QUESTIONS ON THE APPLICATION OF TABLES, FORMULAS AND GENERAL DATA IN MACHINERY'S HANDBOOK, SELECTED ESPECIALLY FOR ENGINEERING AND TRADE SCHOOLS, APPRENTICESHIP AND HOME-STUDY COURSES, TO INSURE THE MOST EFFECTIVE USE OF THE HANDBOOK AND A THOROUGH KNOWLEDGE OF ITS CONTENTS

By

JOHN M. AMISS

AND

FRANKLIN D. JONES

1971 EDITION

INDUSTRIAL PRESS INC.
200 Madison Avenue, New York, N. Y. 10016

THE MACHINERY PUBLISHING CO., LTD.
New England Street, Brighton 1, England

Library of Congress Catalog Card Number: 70-163896

ISBN 0-8311-1079-1

THE PURPOSE OF THIS BOOK

An engineering handbook now is an essential part of the equipment of practically all engineers, machine designers, draftsmen, and skilled mechanics in machine shops and tool-rooms. The daily use of such a book, with its various tables and general data saves a lot of time and labor. But to obtain the full value of any handbook, the user must know enough about the contents to apply the tables, formulas, and other data, whenever they can be used to advantage.

One purpose of this book, which is based upon MACHINERY'S HANDBOOK, is to show by examples, solutions and test questions, typical applications of handbook matter in both drafting rooms and machine shops. Another function is to familiarize engineering students or other handbook users with the contents. A third object is to provide test questions and drill work that will enable the handbook user, through practice, to obtain the required information quickly.

MACHINERY'S HANDBOOK, in common with all other handbooks, presents information in condensed form so that a large variety of subjects can be covered in a single volume. Because of this condensed treatment, any engineering handbook must be primarily a work of reference rather than a text-book, and the practical application of some parts will not always be apparent, especially to those who have had little experience in engineering work. The questions and examples in this book are intended not only to supplement some of the HANDBOOK material, but to stimulate interest both in those parts that are used frequently and in the more special sections that may be very valuable even though seldom required.

A thorough working knowledge of an engineering handbook is certain to prove valuable to everyone engaged in the design or manufacture of mechanical apparatus, and this collection of problems and questions is intended to assist students and the younger engineers in acquiring such knowledge. Much of the material in this book has been used effectively by the industrial training department of the Chrysler Corporation and also by the Detroit Coöperative Trade School.

ACKNOWLEDGMENTS

A revision of Section 16, Strength of Materials, has been made by E. T. Fortini to bring it into agreement with the revised section in Machinery's Handbook, 18th Edition.

Henry H. Ryffel, who served as editor of The Use of Handbook Tables and Formulas from 1954 to 1964, was responsible for the revision and expansion of Section 18, Problems in Designing and Cutting Gears.

CONTENTS

The Use of Handbook Tables and Formulas contains numerous references to subjects in MACHINERY'S HANDBOOK. The handbook page numbers accompanying these references apply to the 19th edition.

THE USE OF HANDBOOK
TABLES AND FORMULAS

SECTION 1

TABLES OF POWERS, ROOTS AND RECIPROCALS OF NUMBERS

MACHINERY'S HANDBOOK Pages 1 to 54

Determining the powers and roots of numbers, especially the larger numbers, would require considerable time by ordinary mathematical processes and besides there is a chance of error. This explains why it is the general practice in engineering work to obtain the powers and roots of numbers directly from handbook tables whenever possible. These and other mathematical tables found in MACHINERY'S HANDBOOK have been checked and re-checked so frequently by many thousands of handbook users, that accuracy is assured. (The powers and roots of numbers are explained on Handbook page 106.)

Although the tables of powers and roots include whole numbers up to 2000 and are supplemented by tables of decimals and mixed numbers, it will often be necessary to determine the power or the root of a number, the exact equivalent of which cannot be found in the tables. In such cases it may be and often is possible to obtain values sufficiently accurate for practical work, but even when this is not feasible the tables are valuable as a means of checking results obtained by mathematical processes.

Example 1: — $\sqrt{157.5} = ?$

First determine the *approximate* value of the square root. The square root of 157 equals 12.53 and 158, 12.5698; therefore, it is evident that the square root of 157.5 is between these values. The *square* of 1255 is 1,575,025; hence, the square root of 157.5 is 12.55.

Example 2: — The expression $\sqrt{8.71^3}$ means that the square root of the cube of 8.71 is required. This root can be found with a fair degree of accuracy by means of the table.

The cube of 871 = 660,776,311. As there are two decimal places in 8.71 there will be $3 \times 2 = 6$ decimal places in its cube. Therefore, $8.71^3 = 660.776311$.

The next step is to find the square root of 660.776311. By referring to the square roots 660 and 661, we know that the required root lies between 25.6905 and 25.7099. The square of 257 is 66049; hence, the square root of the given number is, as nearly as it can be determined by the direct application of the table, 25.7. The exact square root to four decimal places is 25.7056.

Reverse Method of Using Tables of Powers and Roots of Numbers. — The tables, Handbook pages 1 to 41, inclusive, give the powers and roots of numbers up to 2000. As the numbers in the second column are the *squares* of the numbers in the first, it follows that the numbers in the first column are the *square roots* of the numbers in the second column. For example, 16 is the square of 4, and 4 is the square root of 16. Similarly the numbers in the first column are the *cube roots* of the numbers or *cubes* in the third column. Thus, 64 is the cube of 4, and 4 is the cube root of 64. This reverse method of using the tables greatly increases their value.

Square Roots of Numbers by Reverse Method of Using Tables. — Frequently it is necessary to find the root of a number that is not given exactly in the table but the roots in such cases often can be determined accurately enough for practical purposes.

Example 3: — Find the square root of 9253 using the table. The table shows that 9216 is the square of 96; hence it is evident that the square root of the given number is a little over 96.

In the column of squares of numbers find a number the first four figures of which are nearest to the four figures in the given number. Thus, on Handbook page 21 we find in the column of squares the number 925444. The first four figures are within one of equalling the given number and this is the square of a number beginning with the figures 9 and 6. Therefore the square root of 9253 is 96.2 nearly. The exact square root of 9253 is 96.192 so that the result obtained by the table is nearly correct.

Example 4: — $\sqrt{9467.27} = ?$

The approximate value of the square root is 97 because the table shows that 9409 is the *square* of 97, or, that 97 is the *square root* of 9409. In the column of squares locate, if possible, a number containing the figures in the given number. The square of 973 is 946,729; therefore, the square root of 9467.27 is 97.3 as nearly as can be determined by the table and accurately enough for most purposes.

Finding Cube Root of Number by Reverse Method. — The indirect use of the table for determining cube roots is similar in principle to that just described for square roots.

Example 5: —Find the cube root of 676836.

On Handbook page 39 the cube of 1891 begins with the figures 676 but the column of cubes on page 3 shows that the cube root is between 87 and 88. On page 19 the cube of 878 is found to be 676,836,152. The first six figures coincide exactly with those in the given number; hence, it is evident that the cube root is 87.8.

Finding the Roots of Decimals. — In using the table of powers and roots for determining the roots of decimals not directly included in the table, consider the decimal the same as though it were a whole number and use the table on Handbook page 1 as a guide for determining the approximate value of the root. The procedure will be illustrated by examples.

Example 6: — $\sqrt{0.0376} = ?$

The approximate value of the root is determined first by referring to the table on Handbook page 1 which shows that the root is between 0.17 and 0.20. The square root of 376 is 19.3907; hence, the square root of 0.0376 is 0.1939.

Example 7: — $\sqrt{0.037635} = ?$

As in the preceding example, the approximate square root is shown by the table on Handbook page 1 to be between 0.17 and 0.20. As the number 37635 is beyond the range of the numbers in the first column of the table, locate in the column of squares a number the first five figures of which are nearest to 37635. On Handbook page 14 will be found the square 376996, but this is opposite number 614 which does not contain the figures of the root as the latter is between 0.17 and 0.20. On page 40 will be found the square 3763600 opposite the number 1940; therefore the square root of the given number is approximately 0.194.

Example 8: — $\sqrt{0.643212} = ?$

The square of 802 is 643204 which is very close to the given number if we disregard the decimal point. The table on Handbook page 1 shows that the square root is slightly over 0.80; therefore, the square root is 0.802.

Example 9: — $\sqrt[3]{0.29539}$ = ?

The cube of 666 is 295,408,296; hence, the cube root is 0.666 very nearly. The errors obtained by this reverse method of using the table are in many cases very slight. In this case the exact root to five decimal places is 0.66599.

Roots of Mixed Numbers. — In using the table to determine the roots of mixed numbers consisting of a whole number and a common fraction, the latter is first changed to a decimal by referring to the table of decimal equivalents on Handbook page 2350, then the procedure is as previously described.

Example 10: — $\sqrt[3]{5592\frac{23}{64}} = \sqrt[3]{5,592.359,375}$

By referring to the column of cubes on Handbook page 2, it will be seen that the cube root of the given number is somewhere between 17 and 18 because the given number is between the cubes 4913 and 5832. The cube of 1775 = 5,592,359,375; therefore, the cube root of the number given in the example is 17.75. Ordinarily, numbers cannot be found in the column of cubes which exactly coincide with all figures in the given number but this method of using the table often is valuable as a check even when the root thus determined is not sufficiently accurate for use.

Obtaining Powers of Fractional Numbers by the Reverse Method. — By the reverse method of using the table, the squares or cubes of fractional numbers may be found; and this is possible because the numbers in the first column are the squares and cubes of the numbers in the fourth and fifth columns, respectively.

Example 11: — Find the square of 4.471, using the Handbook table.

Turning to Handbook page 10, the square of 447 lies between 190,000 and 200,000 so that, roughly, the square of 4.471 is between 19 and 20. On page 41, fourth column, the *square root* of 1999 is 44.7102; or, the *square* of 44.7102 is 1999. Therefore, the square of 4.471 is very nearly 19.99, the square obtained by actual calculation being 19.989841.

Squares of Numbers Ending with 5. — The tables of squares may also be used to obtain quickly the squares of numbers above the range of the table, consisting of four or five figures, the last of which is 5. Assume, for instance, that it is necessary to find the square of 319.5.

Consider 319.5 as the arithmetical mean of 319 and 320; then from Handbook page 8, obtain the values for the squares of these two numbers, add them together and divide by 2. Next, in the quotient place the figure 2 between the last figure on the right-hand side and the one preceding it. The last figure will always be 5 because the sum of the squares of two consecutive numbers is always an odd number, and dividing any odd number by 2 obviously makes the last figure of the quotient 5. To illustrate:

$$319^2 = 101761$$
$$320^2 = 102400$$

Adding = 204161

Dividing by 2,

$$204161 \div 2 = 1020805$$

Placing the figure 2 at the left-hand side of the last figure, we have 10208025. Next, placing the decimal point according to the common rules of multiplication, $319.5^2 = 102080.25$. This result can be readily proved by squaring the number by multiplication. Similarly, $31.95^2 = 1020.8025$ and $3.195^2 = 10.208025$.

That this method is especially useful in finding the square of numbers having a decimal fraction equivalent to eighths, quarters, and sixteenths inch will be apparent from the two following examples:

Example 12: — Square 12.125

$$1212^2 = 1468944$$
$$1213^2 = 1471369$$

Adding = 2940313

Dividing by 2,

$$2940313 \div 2 = 14701565$$

Placing figure 2 at the left of the last figure, we have 147015625. Finally locating the decimal point properly,

$$12.125^2 = 147.015625$$

Example 13: — Square 1.1875

$$1187^2 = 1408969$$
$$1188^2 = 1411344$$

$$\text{Adding} = 2820313$$

Dividing by 2,

$$2820313 \div 2 = 14101565$$

Placing figure 2 at the left of the last figure we have 141015625.
Then

$$1.1875^2 = 1.41015625$$
$$11.875^2 = 141.015625$$
$$118.75^2 = 14101.5625$$

Reciprocals of Numbers. Pages 2 to 41. — By the use of reciprocals, a tedious division problem may be converted into a simple problem of multiplication. The definition of a reciprocal and an explanation of its use may be found on Handbook page 106 The use of reciprocals also is shown by the following example:

Example 14: — A gallon equals what part of a cubic foot?

1 gallon = 231 cubic inches. The reciprocal of 1728 = 0.0005787.

1 cubic foot = 1728 cubic inches. $\frac{231}{1728}$ = 231 × 0.0005787 = 0.13368.

PRACTICE EXERCISES FOR SECTION 1

For answers to all practice exercise problems or questions see
Section 20

1. Find the square root and cube root of 49; 75; 0.84; 0.27.
2. Square $3\frac{3}{16}$; $5\frac{31}{32}$; $11\frac{7}{8}$; 4.1875; 7.4375.
3. Find the square, cube, square root and cube root of 36; 73; 197; 1291; $\frac{1}{8}$; $\frac{5}{16}$; 0.3125; 0.609375.
4. Find the square and cube and the square root and cube root of $36\frac{1}{4}$; $\frac{31}{64}$; $\frac{29}{32}$.
5. Find the square root and cube root of $\frac{23}{32}$; $\frac{57}{64}$; 0.984375; 0.09.
6. Find the cube root of 43,243,551; 7,809,531,904.
7. Find the reciprocal of 923; 1728; 231.
8. Solve 984 ÷ 423 by the use of reciprocals.
9. 14 cubic inches equals what decimal of a cubic foot? (Use reciprocals).
10. $\frac{5}{8}$ inch equals what decimal of a yard?

SECTION 2

DIMENSIONS AND AREAS OF CIRCLES

Handbook Pages 55 to 71

Circumferences of circles are used in calculating speeds of machine parts, including: drills, reamers, cutters, grinding wheels, gears and pulleys. These speeds are variously referred to as: surface speed, circumferential speed, and peripheral speed; meaning in each case the distance a point on the surface or circumference would travel per minute. This distance is usually expressed as feet per minute. Many formulas involve the use of multiples of 3.1416 (Ratio between the circumference and the diameter of a circle). Such formulas may be applied more rapidly by finding the appropriate circumference, using the multiple of pi as diameter. Circumferences are also required in calculating the circular pitch of gears, laying out involute curves, finding the lengths of arcs, and in solving many geometrical problems. Letters from the Greek alphabet are frequently used to designate angles, and the Greek letter π (pi) is always used to indicate the ratio between the circumference and the diameter of a circle; π, therefore, is always, in mathematical expressions, equal to 3.1416.

Example 1: — Use of table of circumferences and areas. Find the circumference and area of a circle whose diameter is $9\frac{5}{8}$.

On Handbook page 55 find $9\frac{5}{8}$ in diameter column; opposite $9\frac{5}{8}$ find 30.2378 for circumference and 72.760 for area.

Example 2: — The area of a cylinder (excluding the ends) equals $2\pi rh$, where r equals the radius and h equals the height of the cylinder.

Find the area of a cylinder whose height is $4\frac{1}{4}$ and radius $2\frac{1}{2}$. Area = $2\pi rh = 2\pi \times 2\frac{1}{2} \times 4\frac{1}{4} = 21\frac{1}{4}\pi$.

On Handbook page 56 opposite the diameter $21\frac{1}{4}$ find in the circumference column 66.7588. (If the dimensions of the cylinder are in inches, the area will be in square inches; and, if the dimensions are in feet, the area will be in square feet.)

Example 3: — If the diameter of a circle is 6.29, what is the circumference?

7

First find the circumference for 629 which is 1976.06; move the decimal point two more places to the left (because 6.29 has two decimal places) obtaining 19.7606, which is the circumference for a diameter of 6.29.

Fractions of 3.1416 (π, or pi) — Handbook page 51. — The ratio of the circumference of a circle to its diameter is commonly given as 3.1416. This ratio, however, cannot be expressed by an exact arithmetical figure because it is an incommensurable number or a number having an infinite number of decimal places. The ratio has been calculated to as many as 707 decimal places. The first twenty-five decimals are as follows:

$$3.1415926535897932384626434$$

For practical use, the first four decimals only are required. As the fifth decimal is "9," the fourth decimal is raised to "6" when only four decimals are used. Fractional approximations that give very close results are to assume π equal to $\frac{22}{7}$ or $\frac{355}{113}$.

The table on Handbook page 51 gives various quotients obtained by dividing π by whole numbers ranging from 1 to 100. Thus, if the divisor is 21, the quotient equals 0.14960.

Example 4: — What is the circumference of a circle if the diameter equals $5\frac{2}{89}$ inches; or $5\frac{2}{89} \pi = ?$

The table on Handbook page 55 shows that the circumference for a diameter of 5 equals 15.7080. The table on page 51 shows that $\pi \div 89 = 0.03530$ which, multiplied by $2 = 0.0706$ and $15.7080 + 0.0706 = 15.7786$ inches.

PRACTICE EXERCISES FOR SECTION 2

For answers to all practice exercise problems or questions
see Section 20

1. Find areas of circles having diameters of: $3\frac{1}{16}$; $12\frac{7}{8}$; 936.
2. Find diameters and areas of circles if their circumferences equal: 10 ft. 8 in.; 21 ft. $2\frac{1}{2}$ in.
3. Find the diameters if the circumferences equal: 21; 186.
4. $\frac{\pi}{17} = ?$ $\frac{\pi}{69} = ?$
5. Circular areas are: 82.516; 38.485; 352565. Find their respective diameters.

6. If circumferences are: 2739.47; 265.857 and 36.5210, find their corresponding diameters and areas.

7. Find the area of the cylindrical surface of a cylinder if its diameter is $7\frac{1}{4}$ inches and height $3\frac{3}{4}$ inches.

8. The drilling speed for cast iron is assumed to be 70 feet per minute. Find the time required to drill two holes in each of 500 castings if each hole has a diameter of $\frac{3}{4}$ inch and is 1 inch deep. Use 0.010 inch feed and allow one-fourth minute per hole for set-up.

9. Find the weight of a cast-iron column 10 inches in diameter and 10 feet high. Cast iron weighs 0.26 pound per cubic inch.

10. If machine steel has a tensile strength of 55,000 pounds per square inch, what should be the diameter of a rod to support 36,000 pounds if the safe working stress is assumed to be one-fifth of the tensile strength?

11. Moving the circumference of a 16-inch automobile flywheel two inches, moves the camshaft through how many degrees? (The camshaft rotates at one-half the flywheel speed.)

12. Where an auxiliary rotary table is not available, what method is followed in locating equally-spaced holes around a circle for jig-boring?

SECTION 3

CHORDAL DIMENSIONS, SEGMENTS, SPHERES AND SPHERICAL SEGMENTS

Handbook Pages 72, 75 and 82

A chord of a circle is the distance along a straight line from one point on the circumference to any other point. A segment of a circle is that part or area between a chord and the arc it intercepts. The lengths of chords and the dimensions and areas of segments are often required in mechanical work.

Lengths of Chords.—The table of chords, Handbook page 75, can be applied to a circle of any diameter as explained and illustrated by examples on page 74. This table is given to six decimal places so that it can be used in connection with precision tool work.

Example 1:—A circle has 56 equal divisions and the chordal distance from one division to the next is 2.156 inches. What is the diameter of the circle?

The chordal length in the table for 56 divisions and a diameter of 1 equals 0.05607; therefore, in this case

$$2.156 = 0.05607 \times \text{diameter}$$

$$\text{Diameter} = \frac{2.156}{0.05607} \quad 38.452 \text{ inches}$$

Example 2:—A drill jig is to have 8 holes equally spaced around a circle 6 inches in diameter. How can the chordal distance between adjacent holes be determined provided table, Handbook page 75, is not available?

One-half the angle between the radial center-lines of adjacent holes = 180 ÷ number of holes. If the sine of this angle is multiplied by the diameter of the circle, the product equals the chordal distance. In this example we have 180 ÷ 8 = 22.5 degrees. The sine of 22.5 degrees (see page 201) is 0.38268; hence, the chordal distance = 0.38268 × 6 = 2.296 inches. The result is the same as would be obtained with the table on Handbook page 75 because the figures in the column "Length of Chord" represent the sines of angles equivalent to 180 divided by the different numbers of spaces.

Use of the Table of Segments of Circles — Handbook pages 72 and 73. — This table is of the unit type in that the values all apply to a radius of 1. As explained above the table, the value for any other radius can be obtained by multiplying the figures in the table by the given radius, except in the case of areas when the *square* of the given radius is used. Thus, the unit type of table is universal in its application.

Example 3: — Find the area of a segment of a circle, the center angle of which is 57 degrees and the radius $2\frac{1}{2}$ inches.

First locate 57 degrees in the center angle column; opposite this figure in the area column will be found 0.07808. Since the area is required, this number is multiplied by the square of $2\frac{1}{2}$. Thus,

$$0.07808 \times (2\tfrac{1}{2})^2 = 0.488 \text{ square inch}$$

Example 4: — A cylindrical oil tank is $4\frac{1}{2}$ feet in diameter, 10 feet long, and is in a horizontal position. When the depth of the oil is 3 feet 8 inches, what is the number of gallons of oil?

The total capacity of the tank equals $0.7854 \times 4\frac{1}{2}^2 \times 10$ = 159 cubic feet.

One U. S. gallon equals 0.1337 cubic feet (see Handbook page 2349); hence, the total capacity of the tank equals $159 \div 0.1337$ = 1190 gallons.

The unfilled area at the top of the tank is a segment having a height of 10 inches or $\frac{10}{27}$ (0.37037) of the tank radius. The nearest decimal equivalent to $\frac{10}{27}$ in Column h of the table on pages 72 and 73 is 0.3707; hence, the number of cubic feet in the segment-shaped space $= \dfrac{27^2 \times 0.401 \times 120}{1728} = 20.3$ cubic feet and $20.3 \div 0.1337 = 152$ gallons. Therefore, when the depth of oil is 3 feet 8 inches, there are $1190 - 152 = 1038$ gallons. (See also Handbook page 166 for additional information on the capacity of cylindrical tanks.)

Spheres and Spherical Segments — Handbook pages 82 to 85. — The table, pages 82 to 84, gives the surface areas of spheres in square inches and their volume in cubic inches. This table gives the values directly for diameters ranging from $\frac{1}{64}$ up to 200.

Example 5: — If the diameter of a sphere is $24\frac{5}{8}$ inches what is the volume?

This diameter is not given in the table but the table is based upon the formula:

$$\text{Volume} = 0.5236 \, d^3$$

The table, page 47, shows that the cube of $24\frac{5}{8} = 14932.369$; hence, the volume of this sphere $= 0.5236 \times 14932.369 = 7818.5$ cubic inches.

PRACTICE EXERCISES FOR SECTION 3

For answers to all practice exercise problems or questions
see Section 20

1. Find the lengths of chords when the number of divisions of a circumference and the radii are as follows: 30 and 4; 14 and $2\frac{1}{2}$; 18 and $3\frac{1}{2}$.

2. Find the chordal distance between the graduations for thousandths on the following dial indicators: (a) Starrett has 100 divisions and $1\frac{3}{8}$-inch dial. (b) Brown & Sharpe has 100 divisions and $1\frac{3}{4}$-inch dial. (c) Ames has 50 divisions and $1\frac{5}{8}$-inch dial.

3. The teeth of gears are evenly spaced on the pitch circumference. In making a drawing of a gear, how wide should the dividers be set to space 28 teeth on a 3-inch diameter pitch circle?

4. In a drill jig, 8 holes, each $\frac{1}{2}$ inch diameter, were spaced evenly on a 6-inch diameter circle. To test the accuracy of the jig, plugs were placed in adjacent holes and the distance over the plugs was measured with a micrometer. What should be the micrometer reading?

5. In the preceding problem, what should be the distance over plugs placed in alternate holes?

6. What is the length of the arc of contact of a belt over a pulley 2 feet 3 inches in diameter if the arc of contact is 215 degrees?

7. Find the areas, length of chords and heights, of the following segments: (a) radius 2 inches, angle 45 degrees; (b) radius 6 inches, angle 27 degrees.

8. Find the number of gallons of oil in a tank 6 feet in diameter and 12 feet long if the tank is in a horizontal position and the oil measures 2 feet deep.

9. Find the surface area of the following spheres, the diameters of which are: $1\frac{1}{2}$; $3\frac{3}{8}$; 65; $20\frac{3}{4}$.

10. Find the volume of each sphere in the above exercise.

11. The volume of a sphere is 1,802,725 cubic inches. What is its surface area and diameter?

12. Find the volumes of spherical segments having angles and chords as follows: 27 degrees and 17.2 inches; 54 degrees and 8.25 inches.

SECTION 4

FORMULAS AND THEIR REARRANGEMENT

Handbook Page 94

A formula may be defined as a mathematical rule expressed by signs and symbols instead of in actual words. In formulas, letters are used to represent numbers or *quantities*, the term "quantity" being used to designate any number involved in a mathematical process. The use of letters in formulas, in place of the actual numbers, simplifies the solution of problems, and makes it possible to condense into small space the information that otherwise would be imparted by long and cumbersome rules. The figures or values for a given problem are inserted in the formula according to the requirements in each specific case. When the values are thus inserted, in place of the letters, the result or answer is obtained by ordinary arithmetical methods. There are two reasons why a formula is preferable to a rule expressed in words. 1. The formula is more concise, it occupies less space, and it is possible to see at a glance, the whole meaning of the rule laid down. 2. It is easier to remember a brief formula than a long rule, and it is, therefore, of greater value and convenience.

Example 1: — In spur gears, the outside diameter of the gear can be found by adding 2 to the number of teeth, and dividing the sum obtained by the diametral pitch of the gear. This rule can be expressed very simply by a formula. Assume that we write D for the outside diameter of the gear, N for the number of teeth, and P for the diametral pitch. Then the formula would be:

$$D = \frac{N + 2}{P}$$

This formula reads exactly as the rule given above. It says that the outside diameter (D) of the gear equals 2 added to the number of teeth (N), and this sum is divided by the pitch (P).

If the number of teeth in a gear is 16 and the diametral pitch 6, then simply put these figures in the place of N and P in the formula, and find the outside diameter as in ordinary arithmetic.

$$D = \frac{16 + 2}{6} = \frac{18}{6} = 3 \text{ inches}$$

Example 2: — The formula for the horsepower of a steam engine is as follows:

$$H = \frac{P \times L \times A \times N}{33,000}$$

in which H = indicated horsepower of engine;

P = mean effective pressure on piston in pounds per square inch;

L = length of piston stroke in feet;

A = area of piston in square inches;

N = number of strokes of piston per minute.

Assume that $P = 90$, $L = 2$, $A = 320$, and $N = 110$; what would be the horsepower?

If we insert the given values in the formula, we have:

$$H = \frac{90 \times 2 \times 320 \times 110}{33,000} = 192$$

From the examples given, we may formulate the following general rule: *In formulas, each letter stands for a certain dimension or quantity; when using a formula for solving a problem, replace the letters in the formula by the figures given for a certain problem, and find the required answer as in ordinary arithmetic.*

Omitting Multiplication Signs in Formulas. — In formulas, the sign for multiplication ($\times$) is often left out between letters the values of which are to be multiplied. Thus AB means $A \times B$, and the formula

$$H = \frac{P \times L \times A \times N}{33,000} \text{ can also be written } H = \frac{PLAN}{33,000}$$

If $A = 3$, and $B = 5$, then: $AB = A \times B = 3 \times 5 = 15$.

It is only the multiplication sign ($\times$) that can be thus left out between the symbols or letters in a formula. All other signs must be indicated the same as in arithmetic. The multiplication sign can never be left out between two figures: 35 always means thirty-five, and "three times five" must be written 3×5; but "three times A" may be written $3A$. As a general rule the figure in an expression such as "$3A$" is written first, and is known as the *coefficient* of A. If the letter is written first, the multiplication sign is not left out, but the expression is written "$A \times 3$."

Rearrangement of Formulas. — A formula can be rearranged or "transposed" to determine the values represented by different letters of the formula. To illustrate by a simple example, the formula for determining the speed (s) of a driven pulley when its diameter (d), and the diameter (D) and speed (S) of the driving pulley are known, is as follows: $s = \dfrac{S \times D}{d}$. If the speed of the driven pulley is known and the problem is to find its diameter or the value of d instead of s, this formula can be rearranged or changed. Thus: $d = \dfrac{S \times D}{s}$.

Rearranging a formula in this way is governed by four general rules.

Rule 1. An independent term preceded by a plus sign ($+$) may be transposed to the other side of the equals sign ($=$) if the plus sign is changed to a minus sign ($-$).

Rule 2. An independent term preceded by a minus sign may be transposed to the other side of the equals sign if the minus sign is changed to a plus sign.

As an illustration of these rules, if $A = B - C$, then $C = B - A$, and if $A = C + D - B$, then $B = C + D - A$. That the foregoing is correct may be proved by substituting numerical values for the different letters and then transposing them as shown.

Rule 3. A term which multiplies all the other terms on one side of the equals sign may be moved to the other side, if it is made to divide all the terms on that side.

As an illustration of this rule, if $A = BCD$, then $\dfrac{A}{BC} = D$ or according to the common arrangement $D = \dfrac{A}{BC}$. Suppose, in the preceding formula, that $B = 10$, $C = 5$, and $D = 3$; then $A = 10 \times 5 \times 3 = 150$, and $\dfrac{150}{10 \times 5} = 3$.

Rule 4. A term which divides all the other terms on one side of the equals sign may be moved to the other side, if it is made to multiply all the terms on that side.

To illustrate, if $s = \dfrac{SD}{d}$, then $sd = SD$, and, according to Rule 3 $d = \dfrac{SD}{s}$. This formula may also be rearranged for determining the values of S and D; thus $\dfrac{ds}{D} = S$, and $\dfrac{ds}{S} = D$.

If, in the rearrangement of formulas, minus signs precede quantities, the signs may be changed to obtain positive rather than minus quantities. All the signs on both sides of the equals sign or on both sides of the equation may be changed. For example, if $-2A = -B + C$, then $2A = B - C$. The same result would be obtained by placing all the terms on the opposite side of the equals sign which involves changing signs. For instance, if $-2A = -B + C$, then $B - C = 2A$.

Fundamental Laws Governing Rearrangement. — After a few fundamental laws which govern any formula or equation are understood, its solution usually is very simple. An equation states that one quantity equals another quantity. So long as both parts of the equation are treated exactly alike the values remain equal. Thus, in the equation $A = \frac{1}{2}ab$, which states that the area A of a triangle equals one half the product of the base a times the altitude b, each side of the equation would remain equal if we added the same amount. $A + 6 = \frac{1}{2}ab + 6$; or we could subtract an equal amount from both sides: $A - 8 = \frac{1}{2}ab - 8$; or multiply both parts by the same number: $7A = 7(\frac{1}{2}ab)$; or we could divide both parts by the same number and we would still have a true equation.

One formula for the total area T of a cylinder is: $T = 2\pi r^2 + 2\pi rh$, where r = the radius and h = the height of the cylinder. Suppose we want to solve this equation for h. $2\pi rh + 2\pi r^2 = T$. Transposing the part which does not contain h to the other side by changing its sign, we get: $2\pi rh = T - 2\pi r^2$. In order to obtain h, we can divide both sides of the equation by any quantity which will leave h on the left-hand side thus:

$$\frac{2\pi rh}{2\pi r} = \frac{T - 2\pi r^2}{2\pi r}$$

It is clear that in the left-hand member, the $2\pi r$ will cancel out, leaving: $h = \frac{T - 2\pi r^2}{2\pi r}$. The expression $2\pi r$ in the right-hand member cannot be cancelled because it is not an independent factor since the numerator equals the difference between T and $2\pi r^2$.

Example 3: — Rearrange the formula for a trapezoid (Handbook page 150) to obtain h.

$$A = \frac{(a+b)h}{2}$$

$$2A = (a + b)h \quad \text{(multiply both members by 2)}$$

$$(a + b)h = 2A \quad \text{(transpose both members so as to get the multiple of } h \text{ on the left-hand side)}$$

$$\frac{(a + b)h}{a + b} = \frac{2A}{a + b} \quad \text{(divide both members by } a + b\text{)}$$

$$h = \frac{2A}{a + b} \quad \text{(cancel } a + b \text{ from the left-hand member)}$$

Example 4: — The formula for determining the radius of a sphere (Handbook page 160) is as follows:

$$r = \sqrt[3]{\frac{3V}{4\pi}}$$

Rearrange to obtain a formula for finding the volume V.

$$r^3 = \frac{3V}{4\pi} \quad \text{(cube each side)}$$

$$4\pi r^3 = 3V \quad \text{(multiply each side by } 4\pi\text{)}$$

$$3V = 4\pi r^3 \quad \text{(transpose both members)}$$

$$\frac{3V}{3} = \frac{4\pi r^3}{3} \quad \text{(divide each side by 3)}$$

$$V = \frac{4\pi r^3}{3} \quad \text{(cancel 3 from left-hand member)}$$

The procedure has been shown in detail to indicate the underlying principles involved. The rearrangement could be simplified somewhat by direct application of the rules previously given. To illustrate:

$$r^3 = \frac{3V}{4\pi} \quad \text{(cube each side)}$$

$$4\pi r^3 = 3V \quad \text{(applying } Rule \ 4 \text{ move } 4\pi \text{ to left-hand side)}$$

$$\frac{4\pi r^3}{3} = V \quad \text{(move 3 to left-hand side — } Rule \ 3\text{)}$$

This final equation would, of course, be reversed to locate V at the left of the equals sign as this is the usual position for whatever letter represents the quantity or value to be determined.

Example 5: — It is required to determine the diameter of cylinder and length of stroke of a steam engine to deliver 150 horsepower. The mean effective steam pressure is 75 pounds; the number of strokes per minute is 120. The length of the stroke is to be 1.4 times the diameter of the cylinder.

First insert in the horsepower formula (Example 2) the known values:

$$150 = \frac{75 \times L \times A \times 120}{33,000} = \frac{3 \times L \times A}{11}$$

The last expression is found by cancellation.

Assume now that the diameter of the cylinder in inches equals D. Then $L = \frac{1.4 D}{12} = 0.117\, D$, according to the requirements in the problem; the divisor 12 is introduced to change the inches to feet, L being in feet in the horsepower formula. The area $A = D^2 \times 0.7854$. If we insert these values in the last expression in our formula, we have:

$$150 = \frac{3 \times 0.117\, D \times 0.7854\, D^2}{11} = \frac{0.2757\, D^3}{11}$$

$$0.2757\, D^3 = 150 \times 11 = 1650$$

$$D^3 = \frac{1650}{0.2757} \qquad D = \sqrt[3]{\frac{1650}{0.2757}} = \sqrt[3]{5984.8} = 18.15$$

Hence, diameter of the cylinder should be about $18\frac{1}{4}$ inches, and the length of the stroke $18.15 \times 1.4 = 25.41$, or, say, $25\frac{1}{2}$ inches.

Solving Equations or Formulas by Trial. — One of the equations used for spiral gear calculations, when the shafts are at right angles, the ratios are unequal and the center distance must be exact, is as follows:

$$R \sec \alpha + \operatorname{cosec} \alpha = \frac{2\, CP_n}{n}$$

In this equation

R = ratio of number of teeth in large gear to number in small gear;

C = exact center distance;

P_n = normal diametral pitch;

n = number of teeth in small gear.

The exact spiral angle α of the large gear is found by trial using the equation just given.

Equations of this form are solved by trial by selecting an angle assumed to be approximately correct, and inserting the secant and cosecant of this angle in the equation, adding the values thus obtained, and comparing the sum with the known value to the right of the equals sign in the equation. An example will show this more clearly. Using the problem given in MACHINERY'S HANDBOOK (page 927) as an example, $R = 3$; $C = 10$; $P_n = 8$; $n = 28$.

Hence, the whole expression

$$\frac{2\,CP_n}{n} = \frac{2 \times 10 \times 8}{28} = 5.714$$

from which it follows that:

$$R \sec \alpha + \operatorname{cosec} \alpha = 5.714$$

In the problem given, the approximate spiral angle required is 45 degrees. The spiral gears, however, would not meet all the conditions given in the problem, if the angle could not be slightly modified. In order to determine whether the angle should be greater or smaller than 45 degrees, insert the values of the secant and cosecant of 45 degrees in the formula. The secant of 45 degrees is 1.4142, and the cosecant, 1.4142. Then,

$$3 \times 1.4142 + 1.4142 = 5.6568$$

The value 5.6568 is too small, as it is less than 5.714, which is the required value. Hence, try 46 degrees. The secant of 46 degrees is 1.4395, and the cosecant, 1.3902. Then,

$$3 \times 1.4395 + 1.3902 = 5.7087$$

Obviously an angle of 46 degrees is too small. Proceed, therefore, to try an angle of 46 degrees, 30 minutes. This angle will be found too great. Similarly 46 degrees, 15 minutes, if tried, will be found too great, and by repeated trials it will finally be found that an angle of 46 degrees, 6 minutes, the secant of which is 1.4422, and the cosecant, 1.3878, meets the requirements. Then,

$$3 \times 1.4422 + 1.3878 = 5.7144$$

which is as close to the required value as necessary.

In general, when an equation must be solved by the trial-and-error method, all the known quantities may be written on the right-hand side of the equal sign, and all the unknown quantities on the left-hand side. A value is assumed for the unknown quantity. This value is substituted in the equation, and all the values thus obtained on the left-hand side are added. In general, if the result is greater than the known values on the right-hand side, the assumed value of the unknown quantity is too great. If the result obtained is smaller than the sum of the known values, the assumed value for the unknown quantity is too small. By thus adjusting the value of the unknown quantity until the left-hand member of the equation with the assumed value of the unknown quantity will just equal the known quantities on the right-hand side of the equal sign, the correct value of the unknown quantity may be determined.

Derivation of Formulas. — Most formulas in engineering handbooks are given without showing how they have been derived or originated, because engineers and designers usually want only the final results; moreover, such derivations would require considerable additional space and they belong in textbooks rather than in handbooks which are primarily works of reference. Although MACHINERY'S HANDBOOK contains thousands of standard and special formulas, it is apparent that no handbook can include every kind of formula, because a great many formulas apply only to local designing

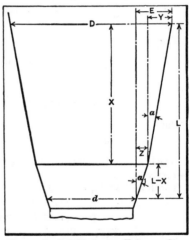

Fig. 1. To find Dimension X from a Given Diameter D to the Intersection of Two Conical Surfaces

or manufacturing problems. Such special formulas are derived by engineers and designers for their own use. The exact methods of deriving formulas are based upon mathematical principles as they are related to the particular factors in each case. A few examples will be given to show how several different types of special formulas have been derived.

Example 6: — The problem is to deduce the general formula

for finding the point of intersection of two tapers with reference to measured diameters on those tapers. In the diagram, Fig. 1,

L = the distance between the two measured diameters, D and d;

X = the required distance from one measured diameter to the intersection of tapers;

a = angle of long taper as measured from center line;

a_1 = angle of short taper as measured from center line.

Then

$$E = \frac{D - d}{2} = Z + Y$$
$$Z = (L - X) \tan a_1$$
$$Y = X \tan a$$

Therefore:

$$\frac{D - d}{2} = (L - X) \tan a_1 + X \tan a$$

and

$$D - d = 2 \tan a_1 (L - X) + 2 X \tan a \qquad (1)$$

But

$$2 \tan a_1 = T_1 \qquad \text{and} \qquad 2 \tan a = T$$

in which T and T_1 represent the long and short tapers per inch, respectively.

Therefore from Equation (1)

$$D - d = T_1(L - X) + TX$$
$$D - d = T_1 L - T_1 X + TX$$
$$X(T_1 - T) = T_1 L - (D - d)$$
$$X = \frac{T_1 L - (D - d)}{T_1 - T}$$

Example 7: — A flywheel is 16 feet in diameter (outside measurement) and the center of its shaft is 3 feet above the floor. Derive a formula for determining how long the hole in the floor must be to permit the flywheel to turn.

The conditions are as represented in Fig. 2. The line AB is the floor level and is a chord of the arc ADB; it is parallel to the horizontal diameter through the center O. CD is the vertical diameter and is perpendicular to AB. It is shown in geometry that the diameter CD bisects the chord AB at the point of intersection E. Now, one of the most useful theorems of geometry is that when a diameter bisects a chord, the product of the two parts of the diameter is equal to the square of one half the chord; in other words, $AE^2 = ED \times EC$. If AB is represented by L and OE by a, $ED = r - a$ and $EC = r + a$, in which $r =$ the radius OC; hence,

$$\left(\frac{L}{2}\right)^2 = (r - a)(r + a) = r^2 - a^2$$

$$\frac{L}{2} = \sqrt{r^2 - a^2}$$

and

$$L = 2\sqrt{r^2 - a^2}$$

Substituting the values given,

$$L = 2\sqrt{8^2 - 3^2}$$
$$= 14.8324 \text{ feet} =$$
14 feet, 10 inches

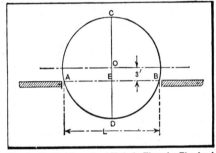

Fig. 2. To find Length of Hole in Floor for Flywheel

The length of the hole should be at least 15 feet, to allow for clearance.

Empirical Formulas.—Many formulas used in engineering calculations can not be established fully by mathematical derivation, but must be based upon actual tests instead of relying upon mere theories or assumptions which might introduce excessive errors. These formulas are known as "empirical formulas." Usually such a formula contains a constant (or constants in some cases) which represents the result of the tests; consequently the value obtained by the formula is consistent with these tests or with actual practice.

A simple example of an empirical formula will be found on Handbook page 1113. This particular formula contains the constant 54,000 which was established by tests, and the formula is used to obtain the breaking load of wrought-iron crane chains to which a factor of safety of 3, 4, or 5 is then applied to obtain the working load. Other examples of empirical formulas will be found on page 428.

On the Handbook page 454 is an example of an empirical formula based upon experiments made with power-transmitting shafts. This formula gives the diameter of shaft required to prevent excessive twisting of shafts.

Parentheses. — Two important rules relating to the use of parentheses are based upon the principles of positive and negative numbers:

1. If a parenthesis is preceded by a $+$ sign, it may be removed, if the terms within the parentheses retain their signs.

$$a + (b - c) = a + b - c$$

2. If a parenthesis is preceded by a $-$ sign, it may be removed, if the signs preceding each of the terms inside of the parentheses are changed ($+$ changed to $-$, and $-$ to $+$). Multiplication and division signs are not affected.

$$a - (b - c) \quad = a - b + c$$
$$a - (-b + c) = a + b - c$$

Knowledge of algebra is not necessary in order to make successful use of formulas of the general type such as are found in engineering handbooks; it is only necessary to thoroughly understand the use of letters or symbols in place of numbers, and to be well versed in the methods, rules, and processes of ordinary arithmetic. Knowledge of algebra becomes necessary only in cases where a general rule or formula which gives the answer to a problem directly is not available. In other words, algebra is useful in *developing* or originating a general rule or formula, but the formula can be *used* without recourse to algebraic processes.

Useful Constants. — Handbook pages 81, 290. — A constant is a value that does not change or is not variable. However, constants at one stage of a mathematical investigation may be variables at another stage, but an *absolute constant* has the same value under all circumstances. The ratio of the circumference to the diameter of a circle, or 3.1416, is a simple example of an absolute constant. In a common formula used for determining the indicated horsepower of a reciprocating steam engine, the product of the mean effective pressure, the length of the stroke in feet, the area of the piston in square inches, and the number of piston strokes per minute is divided by the constant 33,000, which represents the number of foot pounds of work per minute equivalent to one horsepower. Constants occur in many mathematical formulas.

1, "Useful Constants Multiplied and
ains nine constants used in engineering
er part of the table gives the products of
plied by numbers from 2 to 9 and the lower
gives quotients obtained by dividing with num-
9. The upper table on page 81, different prod-
ents, powers and roots of π (3.1416) are given in order
ime in numerical calculations. The same applies to the
rical values for the constant g which represents the accelera-
on due to gravity. (The practical application of this constant is explained in the section on mechanics under "Acceleration of Gravity, g, Used in Mechanics," page 290.) The second half of the table gives constants for weights and volumes of water and air.

Example 9: — If a column of water in a standpipe is 100 feet high, what is the pressure of water per square inch at the bottom of the pipe?

According to one of the constants on page 81 (see upper table), a column of water 1 inch square and 1 foot high weighs 0.4335 pound, so that the pressure at the bottom of this column is 0.4335 pound per square inch. Therefore, if the column of water is 100 feet high, the pressure equals $0.4335 \times 100 = 43.35$ pounds per square inch.

Mathematical Signs and Abbreviations — Handbook page 288. — Every division of mathematics has its traditions, customs and signs, which are frequently of ancient origin. Hence we encounter Greek letters in many problems where it would seem that English letters would do as well or better. Most of the signs on page 288 will be used frequently. They should therefore be understood.

Conversion Tables. — In some cases it may be necessary to convert English units of measurement into metric units and vice versa. The tables provided in the latter portion of the Handbook will be found useful in this connection.

PRACTICE EXERCISES FOR SECTION 4

For answers to all practice exercise problems or questions see Section 20

1. An approximate formula for determining the horsepower H of automobile engines is: $H = \frac{D^2SN}{3}$, where $D =$

diameter of bore, inches; S = length of strok█
N = number of cylinders. Find the horsepow█
following automobile engine: (*a*) bore, $3\frac{1}{2}$█
stroke, $4\frac{1}{4}$ inches; 6 cylinders. (*b*) Using the reci█
of 3, how could this formula be stated?

2. Using the right-angle triangle formula: $C = \sqrt{a^2 +█}$
where a = one side, b = the other side and $C = $ t█
hypotenuse, find the hypotenuse of a right triangl█
whose sides are 16 inches and 63 inches.

3. A formula for finding the blank diameter of a cylindrical
shell is: $D = \sqrt{d \times (d + 4 h)}$, where D = blank diameter; d = diameter of the shell; h = height of the
shell. Find the diameter of the blank to form a cylindrical shell 3 inches diameter and 2 inches high.

4. If D = diagonal of a cube; d = diagonal of face of a cube;
s = side of a cube and V = volume of a cube; then
$d = \sqrt{\dfrac{2 D^2}{3}}$; $s = \sqrt{\dfrac{D^2}{3}}$ and $V = s^3$. Find the side,
volume of a cube, and diagonal of the face of a cube if
the diagonal of the cube is 10.

5. The area of an equilateral triangle equals one fourth of
the side squared times the square root of 3, or $A = \dfrac{S^2}{4} \sqrt{3} = 0.43301 \ S^2$. Find the area of an equilateral
triangle the side of which is 14.5 inches.

6. To what do the constants in the lower table on Handbook page 81 refer?

7. $1728 \times 6 = ?$ $3.1416 \times 8 = ?$ $0.7854 \div 7 = ?$

8. Find the reciprocal of π; $\pi^3 = ?$ $1 \div \sqrt{g} = ?$

9. The formula for the volume of a sphere is: $\dfrac{4 \pi r^3}{3}$ or $\dfrac{\pi d^3}{6}$.

What constants may be used in place of $\dfrac{4 \pi}{3}$ and $\dfrac{\pi}{6}$?

10. The formula for the volume of a solid ring is $2 \pi^2 R r^2$, where
r = radius of cross section and R = radius from the

center of the ring to the center of the cross section. Find the volume of a solid ring made from 2-inch round stock if the mean diameter of the ring is 6 inches.

11. Explain the following signs: $\pm$, $>$, $<$, tan, $\angle$, $\sqrt[4]{}$, log, θ, β, $\sin^{-1} a$, $::$, I.H.P.

12. The area A of a trapezoid (see Handbook page 150) is found by the formula:

$$A = \frac{(a + b)h}{2}$$

Transpose the formula for determining width a.

13. $R = \sqrt{r^2 + \dfrac{s^2}{4}}$; solve for r.

14. $P = 3.1416 \sqrt{2\,(a^2 + b^2)}$; solve for a.

15. $\cos A = \sqrt{1 - \sin^2 A}$; solve for $\sin A$.

16. $\dfrac{a}{\sin A} = \dfrac{b}{\sin B}$; solve for a, b, $\sin A$, $\sin B$.

SECTION 5

LOGARITHMS AND THEIR PRACTICAL APPLICATION

Handbook Pages 120 to 147

The purpose of logarithms is to facilitate and shorten calculations involving multiplication and division, obtaining the powers of numbers and extracting the roots of numbers. By means of logarithms, long multiplication problems become simple addition of logarithms; cumbersome division problems are easily solved by simple subtraction of logarithms; the fourth root or say the 10.4th root of a number can be extracted easily, and any number can be raised to the twelfth power as readily as it can be squared.

In the common or Briggs system of logarithms which is used ordinarily, the base of the logarithms is 10; that is, the logarithm is the *exponent* that would be affixed to 10 to produce the number corresponding to logarithm. To illustrate, by taking simple numbers:

$$\text{Logarithm of} \quad 10 = 1 \text{ because } 10^1 = \quad 10$$
$$\text{Logarithm of} \quad 100 = 2 \text{ because } 10^2 = \quad 100$$
$$\text{Logarithm of } 1000 = 3 \text{ because } 10^3 = 1000$$

In each case it will be seen that the exponent of 10 equals the logarithm of the number. The logarithms of all numbers between 10 and 100 equal 1 plus some fraction. For example: The logarithm of 20 = 1.30103.

The logarithms of all numbers between 100 and 1000 = 2 plus some fraction; between 1000 and 10,000 = 3 plus some fraction, and so on. The tables of logarithms in engineering handbooks give only this fractional part of a logarithm which is called the *mantissa*. The whole number part of a logarithm, which is called the *characteristic*, is not given in the tables because it can easily be determined by simple rules. The logarithm of 350 is 2.54407. The whole number 2 is the characteristic (see rules on Handbook page 120) and the decimal part 0.54407, or the mantissa, is found in the tables.

28

Principles Governing the Application of Logarithms. — When logarithms are used, the product of two numbers can be obtained as follows: Add the logarithms of the two numbers; the sum equals the logarithm of the product. For example: The logarithm of 10 (commonly abbreviated log 10) equals 1; log 100 = 2; 2 + 1 = 3 which is the logarithm of 1000 or the product of 100 × 10.

While logarithms would not be used for such a simple example of multiplication, these particular numbers are employed merely to illustrate the principle involved.

For division by logarithms, subtract the logarithm of the divisor from the logarithm of the dividend to obtain the logarithm of the quotient. To use another simple example, divide 1000 by 100 using logarithms. As the respective logarithms of these numbers are 3 and 2, the difference of 1 equals the logarithm of the quotient 10.

In using logarithms to raise a number to any power, simply multiply the logarithm of the number by the exponent of the number; the product equals the logarithm of the power. To illustrate, find the value of 10^3 using logarithms. The logarithm of 10 = 1 and the exponent is 3; hence, 3 × 1 = 3 = log of 1000; hence 10^3 = 1000.

To extract any root of a number, merely divide the logarithm of this number by the index of the root; the quotient is the logarithm of the root. Thus, to obtain the cube root of 1000, divide 3 (log 1000) by 3 (index of root); the quotient equals 1 which is the logarithm of 10. Therefore

$$\sqrt[3]{1000} = 10$$

Logarithms are of great value in many engineering and shop calculations because they make it possible to solve readily cumbersome and also difficult problems which otherwise would require complicated formulas or higher mathematics. Keep constantly in mind that logarithms are merely exponents. Any number might be the base of a system of logarithms. Thus, if 2 were selected as a base, then the logarithm of 256 would equal 8 because $2^8 = 256$. However, unless otherwise mentioned, the term "logarithm" is used to apply to the common or Briggs system which has 10 for a base.

The tables of common logarithms are found on Handbook pages 126 to 143. The natural logarithms, pages 144 to 147, are based upon the number 2.7183. These logarithms are used in higher mathematics and also in connection with the formula to determine the mean effective pressure of steam in engine cylinders.

Finding the Logarithms of Numbers. — There is nothing complicated about the use of logarithms but a little practice is required to locate readily the logarithm of a given number or to reverse this process and find the number corresponding to a given logarithm. These corresponding numbers are sometimes called "antilogarithms."

Study carefully the rules for finding logarithms given on Handbook page 120. Although the characteristic or whole number part of a logarithm is easily determined, the following table will assist the beginner in memorizing the rules.

Sample Numbers and Their Characteristics

Number	Characteristic	Number	Characteristic
0.008	$\bar{3}$	88	1
0.08	$\bar{2}$	888	2
0.8	$\bar{1}$	8888	3
8.	0	88888	4

Example of the use of the table of numbers and their characteristics: What number corresponds to the log $\bar{2}$.66417? Find 0.66417 in the log tables to correspond to 4615. From the table of characteristics note that a $\bar{2}$ characteristic calls for one zero in front of the first integer; hence point off: 0.04615 as the number corresponding to the log $\bar{2}$.66417.

Example 1: — Find the logarithm of 468.7.

The mantissa of this number is .67089. When there are three whole number places the characteristic is 2; hence the log of 468.7 is 2.67089.

After a little practice with this table, one becomes familiar with the rules governing the characteristic so that reference to this table is no longer necessary.

Obtaining More Accurate Values Than Given Directly by Tables. — The method of using the tables of logarithms to obtain more accurate values than are given directly, by means of interpolation, is explained on Handbook page 125. These instructions should be read carefully in order to understand the procedure in connection with the following example:

Example 2: — $\dfrac{76824 \times 52.076}{435.21}$ = ?

log 76824 = 4.88549	log numerator = 6.60213
log 52.076 = 1.71664	log 435.21 = 2.63870
log numerator = 6.60213	log quotient = 3.96343

The number corresponding to the logarithm 3.96343 is 9192.4. The logarithms just given for the dividend and divisor are obtained by interpolation and in the following manner:

$$\begin{aligned}
\text{Mantissa } 7683 &= .88553 \\
\text{Mantissa } 7682 &= .88547 \\
\hline
\text{Difference} &= .00006
\end{aligned}$$

The number obtained from the table of "proportional parts" (P.P.) is 2.4 and 88547 + 2.4 = 88549.4 so that .88549 is taken as the corrected mantissa of the logarithm of 76824. By again using the table of proportional parts as explained in the Handbook, the corrected mantissas are found for the logarithms of 52.076 and 435.21.

After obtaining the logarithm of the quotient which is 3.96343, the table of proportional parts is again used to determine the corresponding number more accurately than would be possible otherwise. The mantissa .96343 (see Handbook page 142) lies between .96341 and .96346.

$$\begin{aligned}
.96346 - .96341 &= .00005 \\
.96343 - .96341 &= .00002
\end{aligned}$$

Note that the second line gives the difference between the mantissa of the quotient and the nearest smaller mantissa in the Handbook table. In the proportional parts table headed 5 and opposite 2 in the right-hand column, we find 4 in the left-hand column and this is the fifth figure in the number required. Affixing this to 9192 for an additional place, we get 91924. Since the characteristic is 3, the answer is 9192.4.

Changing Form of Logarithm having Negative Characteristic. — The characteristic is frequently rearranged for easier manipu-

lation.   Note that 8 — 8 is the same as 0.   Hence the log of 4.56
could be stated: 0.65896 or 8.65896 — 8.   The log of 0.075 =
$\overline{2}$.87506 or 8.87506 — 10 or 7.87506 — 9.   Any similar arrangement could be made, as determined by ease in multiplication or
division.

Example 3: — $\sqrt[3]{0.47}$ = ?

log 0.47 = $\overline{1}$.67210   or   8.67210 — 9

log $\sqrt[3]{0.47}$ = (8.67210 — 9) ÷ 3 = 2.89070 — 3 or $\overline{1}$.89070.

(8 — 9 was chosen because 3 will divide evenly into 9.   11 — 12
or 5 — 6 could have been used as well.)   (Refer also to Example
2 on Handbook page 124.   The procedure differs from that just
described but the same result is obtained.)

To find the number corresponding to $\overline{1}$.89070, locate the nearest
mantissa.   This is found in the table and corresponds to 7775.
A $\overline{1}$ characteristic indicates that the decimal point immediately
precedes the first integer; therefore the number equivalent to
the log $\overline{1}$.89070 is 0.7775.

Cologarithms. — The cologarithm of a number is the logarithm of the reciprocal of that number.   "Cologs" have no
properties different from those of ordinary logarithms but they
enable division to be carried out by addition because the addition of a colog is the same as the subtraction of a logarithm.

Example 4: — $\dfrac{742 \times 6.31}{55 \times 0.92}$ = ?

Note that this could be stated: $742 \times 6.31 \times \frac{1}{55} \times \dfrac{1}{0.92}$.

Then the logs of each number could be added because the process
is one of multiplication only.

log $\frac{1}{55}$ can be obtained readily in two ways.

log $\frac{1}{55}$ = log 0.0181818 (see reciprocals); log 0.0181818 =
$\overline{2}$.25964; or log $\frac{1}{55}$ = log 1 — log 55

log 1 = 10.00000 — 10

log 55 =   1.74036

8.25964 — 10

or $\overline{2}$.25964

This number $\overline{2}$.25964 is called the colog of 55; hence to find
the colog of any number, subtract the logarithm of that number
from 10.00000 — 10.

To find the colog of 0.92, subtract log 0.92 (or 1.96379) from 10.00000 — 10, thus:

$$
\begin{aligned}
& 10.00000 - 10 \\
\log 0.92 = \ & \underline{1.96379} \\
\text{colog } 0.92 = \ & 10.03621 - 10 \quad \text{or} \quad 0.03621
\end{aligned}
$$

(In subtracting negative characteristics, change the sign of the lower one and add.)

Another method is to use log 0.92 = 1.96379 or 9.96379 — 10, and proceeding as above:

$$
\begin{aligned}
& 10.00000 - 10 \\
\log 0.92 = \ & \underline{9.96379 - 10} \\
\text{colog } 0.92 = \ & 0.03621
\end{aligned}
$$

Example 4 may then be solved by adding logs thus:

$$
\begin{aligned}
\log 742 &= 2.87040 \\
\log 6.31 &= 0.80003 \\
\text{colog } 55 &= 2.25964 \\
\text{colog } 0.92 &= \underline{0.03621} \\
\log \text{ quotient} &= 1.96628
\end{aligned}
$$

The number corresponding to the logarithm of the quotient = 92.53.

Study the examples given in the handbook. Time and effort can be saved by using the logarithms of fractions on Handbook pages 52 to 54, and the logarithms of values of 3.1416 or π on page 81.

Example 5: — The initial absolute pressure of the steam in a steam engine cylinder is 120 pounds; the length of the stroke is 26 inches; the clearance 1½ inches; and the period of admission, measured from the beginning of the stroke, 8 inches. Find the mean effective pressure.

The mean effective pressure is found by the formula:

$$ p = \frac{P\,(1 + \log_e R)}{R} $$

in which p = mean effective pressure in pounds per square inch;
     P = initial absolute pressure in pounds per square inch;
     R = ratio of expansion, which in turn is found from the formula:

$$ R = \frac{L + C}{l + C} $$

in which L = length of stroke in inches;
l = period of admission in inches;
C = clearance in inches.

The given values are P = 120; L = 26; l = 8; and C = 1½. By inserting the last three values in the formula for R, we have:

$$R = \frac{26 + 1\frac{1}{2}}{8 + 1\frac{1}{2}} = \frac{27.5}{9.5} = 2.89$$

If we now insert the value of P and the found value of R in the formula for p, we have:

$$p = \frac{120\,(1 + \log_e 2.89)}{2.89}$$

The natural logarithm (hyp. log.) must be found from tables. The natural logarithm for 2.89 is 1.0613 (see Handbook page 144). Inserting this value in the formula, we have:

$$p = \frac{120\,(1 + 1.0613)}{2.89} = \frac{120 \times 2.0613}{2.89} = 85.6 \text{ lbs. per square inch}$$

PRACTICE EXERCISES FOR SECTION 5

For answers to all practice exercise problems or questions
see Section 20

1. What are the rules governing the characteristics?

2. Find the mantissa of: 762; 478; 26; 0.0098; 6743; 24.82.

3. What are the characteristics of the numbers just given?

4. What numbers could correspond to the following mantissas: 0.85016; 0.88508; 0.22763?

5. (a) If the characteristic of each of the mantissas just given is 1, what would the corresponding numbers be?
   (b) Using the following characteristics (2,0,3) for each mantissa, find the antilogarithms or corresponding numbers.

6. Log 765.4 = ?   log 87.2 = ?   log .00874 = ?

7. What are the antilogarithms of: 2.89894; 1.24279; 0.18013; 2.68708?

8. Find by interpolation the logarithm of: 75186; 42.037.

9. Find the numbers corresponding to the following logarithms: 1.82997; 0.67712.

10. $(2.71)^5 = ?$ $(4.23)^{2.5} = ?$

11. $\sqrt{97.62} = ?$ $\sqrt[5]{4687} = ?$ $\sqrt[2.3]{44.5} = ?$

12. $\dfrac{62876 \times 54.2 \times 0.0326}{1728 \times 231} = ?$

13. $\left(\tfrac{2}{19}\right)^7 = ?$

14. $(9.16)^{2.47} = ?$

15. $\sqrt[3]{\dfrac{(75)^2 \times (5.23)^{\frac{3}{4}}}{0.00036 \times \sqrt{51.7}}} = ?$

16. The area of a circular sector = 0.008727 ar^2 where a = angle in degrees and r = radius of the circle. Find the area of a circular sector the radius of which is 6.25 inches and the central angle is 42° 15′.

17. The diameter of a lineshaft carrying pulleys may be found from the formula: $d = \sqrt[3]{\dfrac{53.5 \text{ H.P.}}{\text{r.p.m.}}}$. Find the diameter of shafting necessary to transmit 50 H.P. at 250 r.p.m.

18. The horsepower of a steam engine is found from the formula H.P. $= \dfrac{PLAN}{33,000}$, where P = mean effective pressure in pounds per square inch; L = length of stroke in feet; A = area of piston in square inches, and N = number of strokes per minute = revolutions per minute $\times$ 2. Find the horsepower of a steam engine if the pressure is 120 pounds, stroke 18 inches, piston 10 inches in diameter and the number of revolutions per minute is 125.

19. In the tables of logarithms, beginning on Handbook page 126 why are many of the numbers preceded by asterisks (*)?

20. Can the tables of logarithms be used for addition and subtraction?

SECTION 6

DIMENSIONS, AREAS AND VOLUMES OF GEOMETRICAL FIGURES

Handbook Pages 148 to 170

Mensuration treats of the lengths of lines, areas of surfaces, and volumes of various geometrical figures. The formulas given for the solution of different problems are derived from plane and solid geometry. For purposes of shop mathematics, all that is necessary is to select the appropriate figure and use the formula given. Keep in mind the tables which have been studied and use them in the solution of the formulas whenever this can be done to advantage.

Eighteen rules may be developed directly from the table for polygons on Handbook page 170. These rules will enable one to solve very easily, nearly every problem involving a regular polygon. For instance, in the first "A" column at the left, $A = 7.6942\ S^2$ for a decagon; hence in this case $S = \sqrt{A \div 7.6942}$. In the first "R" column, $R = 1.3066\ S$ for an octagon; hence $S = R \div 1.3066$.

The frequent occurrence of such geometrical figures as squares, hexagons, spheres and spherical segments in shop calculations causes the tables on Handbook page 169 dealing with these figures to be very useful.

Example 1: — A rectangle 12 inches long has an area of 120 square inches; what is the length of its diagonal?

The area of a rectangle equals the product of the two sides; hence, the unknown side of this case equals $\frac{120}{12} = 10$ inches.

$$\text{Length of diagonal} = \sqrt{12^2 + 10^2} = \sqrt{244}$$

The table on Handbook page 6 shows that the square root of $244 = 15.6205$.

Example 2: — If the diameter of a sphere, the diameter of the base and the height of a cone, are all equal find the volume of the sphere if the volume of the cone is 250 cubic inches.

The formula on Handbook page 158 for the volume of a cone,

shows that in this case $250 = 0.2618\, d^2 h$ in which d = diameter of cone base and h = vertical height of cone; hence,

$$d^2 = \frac{250}{0.2618\, h}$$

Since in this example d and h are equal,

$$d^3 = \frac{250}{0.2618}$$

and

$$d = \sqrt[3]{\frac{250}{0.2618}} = 9.8474 \text{ inches}$$

Referring to the formula on Handbook page 160, the volume of a sphere $= 0.5236\, d^3 = 0.5236 \times 9.8474^3 = 500$ cubic inches.

In solving the following exercises, first, construct the figure carefully and then apply the formula. Use the examples in the Handbook as models.

PRACTICE EXERCISES FOR SECTION 6

For answers to all practice exercise problems or questions
see Section 20

1. Find the volume of a cylinder having a base radius of 12.5 and a height of 16.3.

2. Find the area of a triangle the sides of which are 12, 14 and 18 inches in length.

3. Find the volume of a torus or circular ring made from $1\frac{1}{2}$-inch round stock, if its outside diameter is 14 inches.

4. A bar of hexagonal screw stock measures 0.750 inch per side. What is the largest diameter that can be turned from this bar?

5. Using the prismoidal formula, (Handbook page 164) find the volume of the frustum of a regular triangular pyramid if its lower base is 6 inches per side; upper base 2 inches per side and height 3 inches. (Use table on page 170 for areas. The side of the midsection equals one-half the sum of one side of the lower base and one side of the upper base.)

6. What is the diameter of a circle the area of which is equivalent to that of a spherical zone whose radius is 4 inches and height 2 inches?

7. Find the volume of steel ball, $\frac{3}{8}$ inches in diameter.

8. What is the length of the side of a cube if the volume equals the volume of a frustum of a pyramid with square bases, 4 inches and 6 inches per side and 3 inches high?

9. Find the volume of a bronze bushing if its inside diameter is 1 inch, outside diameter is $1\frac{1}{2}$ inches, and length is 2 inches.

10. Find the volume of a hollow sphere if its outside diameter is 10 inches, inside diameter 6 inches.

11. Find the area of a polygon of 10 sides, inscribed in a 6-inch circle.

12. What is the radius of a fillet if its chord is 2 inches? What is its area?

13. Find the area of the conical surface and volume of a frustum of a cone if the diameter of its lower base is 3 feet, diameter of upper base 1 foot and height 3 feet.

14. Find the total area of the sides and the volume of a triangular prism 10 feet high, having a base width of 8 feet.

15. The diagonal of a square is 16 inches. What is the length of its side?

16. How many gallons in a barrel having the following dimensions: height $2\frac{1}{2}$ feet; bottom diameter 18 inches; bilge diameter 21 inches? (The sides are bent to the arc of a circle.)

17. Find the area of a sector of circle if the radius is 8 inches and central angle is 32 degrees.

18. Find the height of a cone if its volume is 17.29 cubic inches and the radius of its base is 4 inches.

19. Find the volume of a rectangular pyramid having a base 4 × 5 inches and height 6 inches.

20. Find the distance across the corners of both hexagons and squares when the distance across flats in each case is: $\frac{1}{2}$; $1\frac{5}{8}$; $3\frac{3}{10}$; 5; 8.

21. The diagonals of squares are: 2.0329; 4.6846. Find the length of each side.

22. In measuring the distance over plugs in a die which has six $\frac{3}{4}$-inch holes equally spaced on a circle, what should be the micrometer reading over opposite plugs, if the distance over alternate plugs is $4\frac{1}{2}$ inches?

23. To what diameter should a shaft be turned in order to mill on one end a hexagon 2 inches on a side; an octagon 2 inches on a side?

SECTION 7

GEOMETRICAL PROPOSITIONS AND CONSTRUCTIONS

Handbook Pages 277 to 287

Geometry is that branch of mathematics which deals with the relations of lines, angles, surfaces, and solids. Plane geometry treats of lines, angles, and surfaces in one plane only; and, since this branch of geometry is of especial importance in mechanical work, the various propositions or fundamental principles are given in the Handbook and also various problems or constructions. This information is particularly useful in mechanical drafting and in solving problems in mensuration.

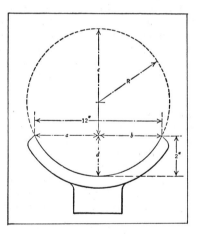

Example 1: — A segment-shaped casting (see illustration) has a chordal length of 12 inches and the height of the chord is two inches; determine by the application of a geometrical principle the radius R of the segment.

This problem may be solved by the application of the first geometrical proposition given on Handbook page 282.

Find Radius R When Length of Chord and Its Height are Given

In this example one chord consists of two sections a and b, each 6 inches long; the other intersecting chord consists of one section d, 2 inches long, and the length of section c is to be determined in order to find radius R.

39

Since $a \times b = c \times d$, it follows that

$$c = \frac{a \times b}{d} = \frac{6 \times 6}{2} = 18 \text{ inches}$$

therefore, $R = \dfrac{c + d}{2} = \dfrac{18 + 2}{2} = 10 \text{ inches}$

In this example one chordal dimension, $c + d =$ the diameter; but, the geometrical principle given in the handbook applies regardless of the relative lengths of the intersecting chords.

Example 2: — The center lines of three holes in a jig plate form a triangle. The angle between two of these intersecting center lines is 52 degrees. Another angle between adjacent center lines is 63 degrees. What is the third angle?

This is an application of the first geometrical principle on Handbook page 277. The unknown angle $= 180 - (63 + 52) = 65$ degrees.

Example 3: — The center lines of four holes in a jig plate form a four-sided figure. Three of the angles between the different intersecting center lines are 63 degrees, 105 degrees, and 58 degrees, respectively. What is the fourth angle?

According to the geometrical principle at the bottom of Handbook page 279, the unknown angle $= 360 - (63 + 105 + 58) = 134$ degrees.

Example 4: — The centers of three holes are located on a circle. The angle between the radial center lines of the first and second holes is 22 degrees, and the center-to-center distance measured along the circle is $2\frac{1}{2}$ inches. The angle between the second and third holes is 44 degrees. What is the center-to-center distance along the circle?

This is an application of the fourth principle on Handbook page 282. Since the lengths of the arcs are proportional to the angles, the center distance between the second and third holes $= \dfrac{44 \times 2\frac{1}{2}}{22} = 5$ inches. (See also rules governing proportion on Handbook page 96.)

The following practice exercises relate to the propositions and constructions given and should be answered without the aid of the handbook:

PRACTICE EXERCISES FOR SECTION 7

For answers to all practice exercise problems or questions
see Section 20

1. If any two angles of a triangle are known, how can the third angle be determined?

2. State three cases where one triangle is equal to another.

3. When are triangles similar?

4. What is the purpose of proving triangles similar?

5. If a triangle is equilateral, what follows?

6. What are the properties of the bisector of any angle of an equilateral triangle?

7. What is an isosceles triangle?

8. How does the size of an angle and the length of a side of a triangle compare?

9. Can you draw a triangle whose sides are 5, 6, and 11 inches?

10. What is the length of the hypotenuse of a right triangle the sides of which are 12 and 16 inches?

11. What is the value of the exterior angle of any triangle?

12. What are the relations of angles formed by two intersecting lines?

13. Draw two intersecting straight lines and a circle tangent to these lines.

14. Construct a right triangle having given the hypotenuse and one side.

15. When are the areas of two parallelograms equal?

16. When are the areas of two triangles equal?

17. If a radius of a circle is perpendicular to a chord, what follows?

18. What is the relation between the radius and tangent of a circle?

19. What line passes through the point of tangency of two tangent circles?

20. What are the attributes of two tangents drawn to a circle from an external point?

21. What is the value of an angle between a tangent and a chord drawn from the point of tangency?

22. Are all angles equal, if their vertices are on the circumference of a circle and they are subtended by the same chord?

23. If two chords intersect within a circle, what is the value of the product of their respective segments?

24. How can a right angle be drawn, using a semi-circle?

25. Upon what does the length of circular arcs in the same circle depend?

26. To what are the circumferences and areas of two circles proportional?

SECTION 8

FUNCTIONS OF ANGLES AND USE OF TABLES
Handbook Pages 171 to 223

The basis of trigonometry is proportion. If the sides of any angle are indefinitely extended and perpendiculars from various points on one side are drawn to intersect the other side, right triangles will be formed and the ratios of the respective sides and hypotenuses will be identical. If the base of the smallest triangle thus formed is 1 inch and the altitude is $\frac{1}{2}$ inch (see Fig. 1) the ratios between these sides is $1 \div \frac{1}{2} = 2$ or $\frac{1}{2} \div 1 = \frac{1}{2}$ depending upon how the ratio is stated. If the next triangle is measured, the ratio between the base and altitude will likewise be either 2 or $\frac{1}{2}$, and this will always be true for any number of triangles, if the angle remains unchanged. For example, $3 \div 1\frac{1}{2} = 2$ and $4\frac{1}{2} \div 2\frac{1}{4} = 2$ or $1\frac{1}{2} \div 3 = \frac{1}{2}$ and $2\frac{1}{4} \div 4\frac{1}{2} = \frac{1}{2}$. This explains why rules can be developed to find the length of any side of a triangle when the angle and one side are known or to find the angle when any two sides are known. Since there are two relations between any two sides of a triangle, there can therefore be a total of six ratios with three sides. These ratios have been defined and explained in the Handbook. Refer to pages 171 and 172 and note explanations of the terms "side adjacent," "side opposite," and hypotenuse.

The abbreviations of the trigonometrical functions begin with a small letter and are not followed by periods.

Functions of Angles and Use of Trigonometrical Tables. — At the bottom of page 171 in the Handbook are given certain rules for determining the functions of angles. These rules, which should be memorized, may also be expressed as simple formulas:

$$\text{sine} = \frac{\text{side opposite}}{\text{hypotenuse}} \qquad \text{cosecant} = \frac{\text{hypotenuse}}{\text{side opposite}}$$

$$\text{cosine} = \frac{\text{side adjacent}}{\text{hypotenuse}} \qquad \text{secant} = \frac{\text{hypotenuse}}{\text{side adjacent}}$$

$$\text{tangent} = \frac{\text{side opposite}}{\text{side adjacent}} \qquad \text{cotangent} = \frac{\text{side adjacent}}{\text{side opposite}}$$

Note that these functions are arranged in pairs to include sine and cosecant; cosine and secant; tangent and cotangent; and each pair consists of a function and its reciprocal. Also, note that the different functions are merely ratios, the sine being the ratio of the *side opposite* to the *hypotenuse;* cosine the ratio

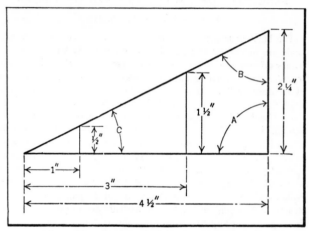

**Fig. 1. For a Given Angle, the Ratio of the Base to the
Altitude is the Same for All Triangle Sizes**

of the *side adjacent* to the *hypotenuse*, etc. The numbers found in the tables of trigonometric functions (Handbook pages 179 to 223) are, therefore, tables of ratios. For example, tan 20° 30′ = 0.37388; this means that in any right triangle having an acute angle of 20° 30′, the side opposite that angle is equal in length to 0.37388 times the length of the side adjacent. Cos 50° 22′ = 0.63787; this means that in any right triangle having an angle of 50° 22′, if the hypotenuse equals a certain length, say 8, the side adjacent to the angle will equal 0.63787 × 8 or 5.10296.

Referring to Fig. 1, tan angle $C = 2\frac{1}{4} \div 4\frac{1}{2} = 1\frac{1}{2} \div 3 = \frac{1}{2} \div 1 = 0.5$; therefore, for this particular angle C, the *side opposite* is always equal to 0.5 times *side adjacent*, thus: $1 \times 0.5 = \frac{1}{2}$; $3 \times 0.5 = 1\frac{1}{2}$, and $4\frac{1}{2} \times 0.5 = 2\frac{1}{4}$. The side opposite angle B equals $4\frac{1}{2}$; hence, tan $B = 4\frac{1}{2} \div 2\frac{1}{4} = 2$. The use of the tables of functions will now be explained.

Finding Angle Equivalent to Given Function. — After determining the tangent of angle C or of angle B, the values of these angles can be determined readily. As tan $C = 0.5$, find the number nearest to this in the tangent column. On Handbook page 205 will be found 0.50003, and the corresponding angle is 26 degrees, 34 minutes. Since angle $A = 90$ degrees, and, as the sum of three angles of a triangle always equals 180 degrees, it is evident that angle $C + B = 90$ degrees; therefore, $B = 90$ degrees minus 26 degrees, 34 minutes = 63 degrees, 26 minutes. The table on Handbook page 205 also shows that the tan 63 degrees, 26 minutes is 1.9998 or 2 within 0.0002.

Note that for angles between 45 and 90 degrees, the table is used by reading from the bottom up, and the minutes are found in the right-hand column as explained on Handbook page 178.

In the foregoing example the tangent is used to determine the unknown angles because the known sides are the *side adjacent* and the *side opposite*, these being the sides required for determining the tangent. If the side adjacent and the length of hypotenuse had been given instead, the unknown angles might have been determined by first finding the cosine because the cosine equals the side adjacent divided by the hypotenuse.

Since the acute angles (like B and C, Fig. 1) of any right triangle must be complementary, the function of any angle equals the co-function of its complement; thus, the sine of angle B = the cosine of angle C; the tangent of angle B = the cotangent of angle C, etc. Thus, tan $B = 4\frac{1}{2} \div 2\frac{1}{4}$ and cotangent C also equals $4\frac{1}{2} \div 2\frac{1}{4}$. The tangent of $20°\ 30' = 0.37388$; which also equals the cotangent of $69°\ 30'$. For this reason, it is only necessary to calculate the trigonometric ratios to $45°$ when making a table of trigonometric functions for angles between $45°$ and $90°$, and this is why the functions of angles between 45 and 90 degrees are located in the table by reading it backwards or in reverse order, as previously mentioned.

Example 1: — Find the tangent of 44 degrees, 59 minutes.

Following instructions given on page 178 of the Handbook, find 44 degrees at the top of page 223, and 59 minutes in the left-hand column headed M; then, opposite 59 and in the column headed "Tan," find the tangent 0.99942.

Example 2: — Find the tangent of 45 degrees, 5 minutes.

The number of degrees is found at the bottom of Handbook page 223 and the number of minutes in the right-hand column headed *M*. Opposite 5, and above "Tan" at the *bottom* of the table, we find the required tangent is 1.0029.

How to Find More Accurate Functions and Angles than are Given ·in Table. — In engineering handbooks, the values of trigonometric functions are usually given to degrees and minutes; hence if the given angle is to degrees, minutes and seconds, the value of the function is determined from the nearest given values by interpolation.

Example 3: — Assume that the sine of 14° 22′ 26″ is to be determined. It is evident that this value lies between the sine of 14° 22′ and the sine of 14° 23′.

Sine 14° 23′ = 0.24841 and sine 14° 22′ = 0.24813. The difference = 0.24841 − 0.24813 = 0.00028. Consider this difference as a whole number (28) and multiply it by a fraction having as its numerator the number of seconds (26) in the given angle, and as its denominator 60 (number of seconds in one minute). Thus $\frac{26}{60} \times 28 = 12$ nearly; hence, by adding 0.00012 to sine of 14° 22′ we find that sine 14° 22′ 26″ = 0.24813 + 0.00012 = 0.24825.

The correction value (represented in this example by 0.00012) is *added* to the function of the *smaller* angle nearest the given angle in dealing with *sines* or *tangents* but this correction value is *subtracted* in dealing with cosines or cotangents.

Example 4: — Find the angle whose cosine is 0.27052.

The table of trigonometric functions shows that the desired angle is between 74° 18′ and 74° 19′ because the cosines of these angles are, respectively, 0.27060 and 0.27032. The difference = 0.27060 − 0.27032 = 0.00028. From the cosine of the *smaller* angle or 0.27060, subtract the given cosine; thus 0.27060 − 0.27052 = 0.00008; hence $\frac{8}{28} \times 60 = 17″$ or the number of seconds to add to the smaller angle to obtain the required angle. Thus the angle for a cosine of 0.27052 is 74° 18′ 17″. Angles corresponding to given sines, tangents, or cotangents may be determined by the same method.

Trigonometric Functions of Angles greater than 90 Degrees. — In obtuse triangles one angle is greater than 90 degrees, and the handbook tables can be used for finding the functions of angles larger than 90 degrees.

The sine of an angle greater than 90 degrees but less than 180 degrees equals the sine of an angle which is the difference between 180 degrees and the given angle.

Example 5: — Find the sine of 118 degrees.

Sin 118° = sin (180° — 118°) = sin 62°. The sine of 118° can also be obtained directly from the table. Angle 118° will be found at the lower left-hand corner of Handbook page 207. For angles between 90 and 135 degrees, the minutes are read in the left-hand column and the functions at the bottom of the table; hence, the sine 118 degrees is found in the column above "Sine" and opposite 0 in the minutes column, the sine being 0.88295. By referring to page 206, it will be seen that this is the sine given for 62 degrees.

The cosine, tangent and cotangent of an angle greater than 90 but less than 180 degrees equals, respectively, the cosine, tangent and cotangent of the difference between 180 degrees and the given angle; but in this case the angular function has a *negative* value and must be preceded by a minus sign.

Example 6: — Find tan 123 degrees, 20 minutes.

Tan 123° 20′ = —tan (180° — 123° 20′) = —tan 56° 40′ = —1.5204. The tangent 123 degrees, 20 minutes is given directly on Handbook page 212, excepting that the minus sign must be added.

Example 7: — Find tangent 150 degrees.

For angles between 135 and 180 degrees, the minutes are found in the right-hand column and the functions at the top of the table.

On Handbook page 208, in the column headed "Tan" and opposite "0" in the right-hand minutes column, is the tangent 0.57735; and, as the value is negative, it is written —0.57735. In the calculation of triangles, it is very important to include the minus sign in connection with the cosines, tangents and cotangents of angles between 90 and 180 degrees. The handbook table on page 178 shows clearly the negative and positive values of different functions and angles.

Use of Functions for Laying Out Angles. — The tables of trigonometric functions may be used for laying out angles accurately either on drawings or in connection with templet work, etc. The following example illustrates the general method of procedure:

Example 8: — Construct or lay out an angle 27 degrees, 29 minutes by using its sine instead of a protractor.

First, draw two lines at right angles and to any convenient length. Find, in the handbook table, the sine of 27 degrees, 29 minutes which equals 0.46149. If there is space enough, lay out the diagram to an enlarged scale in order to obtain greater accuracy. Assume, in this case, that the scale is to be 10 to 1: therefore, multiply the sine of the angle by 10 obtaining 4.6149 or 4$\frac{39}{64}$ very nearly. Set the dividers or the compass to this dimension and with a (Fig. 2) as a center, draw an arc, thus ob-

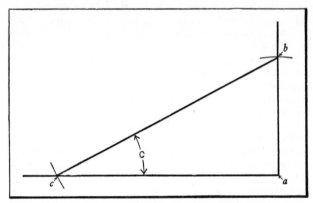

Fig. 2. Method of Laying Out Angle by Using Its Sine

taining one side of the triangle ab. Now set the compass to 10 inches (since the scale is 10 to 1) and with b as the center, describe an arc so as to obtain intersection c. The hypotenuse of the triangle is now drawn through the intersections c and b thus obtaining an angle C of 27 degrees, 29 minutes within fairly close limits. The angle C, laid out in this way, equals 27 degrees, 29 minutes because

$$\frac{\text{Side Opposite}}{\text{Hypotenuse}} = \frac{4.6149}{10} = 0.46149 = \sin 27° 29'$$

Table of Functions used in Conjunction with Formula. — When milling keyways it is often desirable to know the total depth from the outside of the shaft to the bottom of the keyway. With this depth known, the cutter can be fed down to the required depth without taking any measurements other than that indicated by the graduations on the machine. In order to determine the total depth, it is necessary to calculate the height of the arc, which is designated as dimension A in Fig. 3. The formula usually employed to determine A for a given diameter of shaft D and width of key W, is

$$A = \frac{D}{2} - \sqrt{\left(\frac{D}{2}\right)^2 - \left(\frac{W}{2}\right)^2}$$

Another formula which is simpler than the one just given, is used in conjunction with a table of trigonometric functions as arranged in MACHINERY'S HANDBOOK. The formula follows:

$$A = \frac{D}{2} \times \text{versed sine of an angle whose cosecant is } \frac{D}{W}$$

Example 9: — To illustrate the application of this formula, let it be required to find the height A when the shaft diameter D is $\frac{7}{8}$ inch and the width W of the key is $\frac{7}{32}$ inch. Then,

$$\frac{D}{W} = \frac{\frac{7}{8}}{\frac{7}{32}} = \frac{7}{8} \times \frac{32}{7} = 4$$

Now in a table of trigonometric functions, locate the value nearest 4 in the column headed "Cosecants," which is 3.9984. Next,

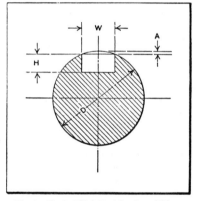

Fig. 3. To find Height A for Arc of Given Radius and Width W

in the column headed "Versed Sine," and on the same line with this cosecant, find the value 0.03178.

Then,

$$A = \frac{D}{2} \times 0.03178 = \frac{7 \times 0.03178}{8 \times 2} = 0.0139 \text{ inch}$$

The total depth of the keyway equals dimension H plus 0.0139 inch.

PRACTICE EXERCISES FOR SECTION 8

For answers to all practice exercise problems or questions
see Section 20

1. How should the tables be used to find angles between 45°
   and 90°?

2. Explain the meaning of sin 30° = 0.50000.

3. Find sin 18° 26' 30"; tan 27° 16' 15"; cos 32° 55' 17".

4. Find the angles which correspond to the following tangents: 0.52035; 0.13025; to the following cosines:
   0.06826; 0.66330.

5. Give two rules for finding *side opposite* a given angle.

6. Give two rules for finding the *side adjacent* a given angle.

7. Explain the following terms: equilateral; isosceles; acute
   angle; obtuse angle; oblique angle.

8. What is meant by complement; side adjacent; side opposite?

9. Can the elements just referred to be used in solving an
   isosceles triangle?

10. Without referring to the Handbook, show the relation-
    ship between the six trigonometric functions and an
    acute angle, using the terms *side opposite*, *side adjacent*
    and *hypotenuse* or abbreviations *SO*, *SA* and *Hyp.*

11. Construct by use of tangents an angle of 42° 20'.

12. Construct by use of sines an angle of 68° 15'.

13. Construct by use of cosines an angle of 55° 5'.

SECTION 9

SOLUTION OF RIGHT-ANGLE TRIANGLES
Handbook Page 174

A thorough knowledge of the solution of triangles or trigonometry is essential in drafting, layout work, bench work, and for convenient and rapid operation of some machine tools. Calculations concerning gears, screw-threads, dovetails, angles, tapers, solution of polygons, gage design, cams, dies, and general inspection work are dependent upon trigonometry. Many geometrical problems may be solved more rapidly by trigonometry than by geometry.

In shop trigonometry it is not necessary to develop and memorize the various rules and formulas; but it is essential that the six trigonometric functions be thoroughly mastered. It is well to remember that a thorough, working knowledge of trigonometry depends upon drill work; hence a large number of problems should be solved.

The various formulas for the solution of right-angle triangles are given on Handbook page 174 and examples showing their application on page 175. These formulas may, of course, be applied to a large variety of practical problems in drafting-rooms, tool-rooms, and machine shops, as indicated by the few examples which follow.

Whenever two sides of a right-angle triangle are given, the third side can always be found by a simple arithmetical calculation, as shown by the third and fourth examples on Handbook page 175. To find the angles, however, it is necessary to use the tables of sines, cosines, tangents and cotangents, and if only one side and one of the acute angles are given, the natural trigonometric functions must be used for finding the lengths of the other sides.

Example 1: — The Jarno taper is 0.600 inch per foot for all numbers. What is the included angle?

As the angle measured from the axis or center line is 0.600 ÷ 2 = 0.300 inch per foot, the tangent of one-half the included angle = 0.300 ÷ 12 = 0.025 = tan 1° 26'; hence the included angle = 2° 52'. A more direct method is to divide the taper per foot by 24 as explained on Handbook page 1559 (see paragraph "To Find Angle for Given Taper per Foot").

Example 2:—Determine the width *W* (see Fig. 1) of a cutter for milling a splined shaft having 6 splines 0.312 inch wide, and a diameter *B* of 1.060 inches.

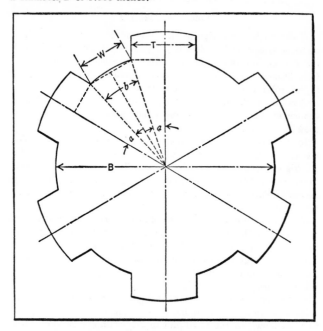

Fig. 1. To find Width *W* of Spline-groove Milling Cutter

This dimension *W* may be computed by using the following formula:

$$W = \sin\left(\frac{\frac{360°}{N} - 2a}{2}\right) \times B$$

in which *N* = number of splines; *B* = diameter of body or of the shafting at the root of the splineway.

Angle *a* must first be computed, as follows:

$$\text{Sin } a = \frac{T}{2} \div \frac{B}{2} \quad \text{or} \quad \text{Sin } a = \frac{T}{B}$$

where T = width of spline; B = diameter at the root of spline-way. In this example

$$\sin a = \frac{0.312}{1.060} = 0.29434 \quad \text{and}$$

$$a = 17° 7'; \quad \text{hence}$$

$$W = \sin\left(\frac{\frac{360°}{6} - 2 \times 17° 7'}{2}\right) \times 1.060 = 0.236 \text{ inch}$$

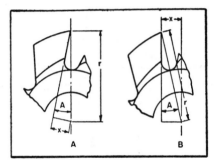

Fig. 2. To Find Horizontal Distance for Positioning Milling Cutter Tooth for Grinding Rake Angle *A*

This formula has also been used frequently in connection with broach design, but it is capable of a more general application. If the splines are to be ground on the sides, suitable deduction must be made from dimension W to leave sufficient stock for grinding.

If the angle *b* is known or is first determined, then

$$W = B \times \sin\frac{b}{2}$$

As there are 6 splines in this example, angle $b = 60° - 2 a = 60° - 34° 14' = 25° 46'$; hence

$$W = 1.060 \times \sin 12° 53' = 1.060 \times 0.22297 = 0.236 \text{ inch}$$

Example 3: — In sharpening the teeth of thread milling cutters, if the teeth have rake, it is necessary to position each tooth for the grinding operation so that the outside tip of the tooth is at horizontal distance x from the vertical center line of the milling cutter as shown in Fig. 2B. What must this distance x be if the outside radius to the tooth tip is r and the rake angle is to be A? What distance x off center must a $4\frac{1}{2}$-inch diameter cutter be set if the teeth are to have a 3-degree rake angle?

In Fig. 2A, it will be seen that, assuming the tooth has been properly sharpened to rake angle A, if a line is drawn extending

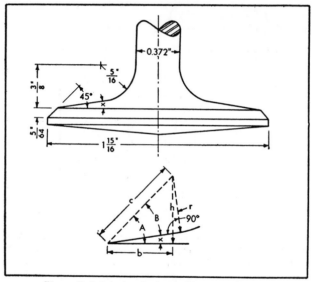

Fig. 3. To find Angle x, having the Dimensions Given on the Upper Diagram

the front edge of the tooth, it will be at a perpendicular distance x from the center of the cutter. Let the cutter now be rotated until the tip of the tooth is at a horizontal distance x from the vertical center line of the cutter as shown in Fig. 2B. It will be noted that an extension of the front edge of the cutter is still at perpendicular distance x from the center of the cutter, indicating that the cutter face is parallel to the vertical center line or is itself vertical, which is the desired position for sharpening, using a vertical wheel. Thus, x is the proper offset distance for grinding the tooth to rake angle A if the radius to the tooth tip is r. Since r is the hypotenuse and x is one side of a right-angled triangle,

$$x = r \sin A$$

For a cutter diameter of $4\frac{1}{2}$ inches and a rake angle of 3 degrees,

$$x = (4.5 \div 2) \sin 3° = 2.25 \times 0.05234$$
$$= 0.118 \text{ inch}$$

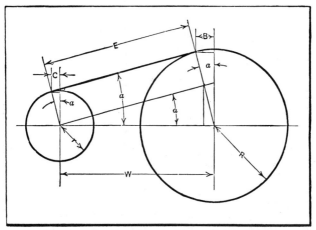

Fig. 4. To find Dimension E or Distance Between Points
of Tangency

Example 4: — Forming tools are to be made for different sizes of poppet valve heads and a general formula is required for finding angle x from dimensions given in Fig. 3.

The values for b, h and r, can be determined easily from the given dimensions. Angle x can then be found in the following manner: Referring to the lower diagram,

$$\tan A = \frac{h}{b} \quad (1) \qquad c = \frac{h}{\sin A} \quad (2)$$

Also,

$$c = \frac{r}{\sin B} = \frac{r}{\sin (A - x)} \quad (3)$$

From Equations (2) and (3) by comparison,

$$\frac{r}{\sin (A - x)} = \frac{h}{\sin A}$$

$$\sin (A - x) = \frac{r \sin A}{h} \qquad (4)$$

From the dimensions given, it is obvious that $b = 0.392125$ inch, $h = 0.375$ inch, and $r = 0.3125$ inch. Substituting these values in Equations (1) and (4) and solving, angle A will be found

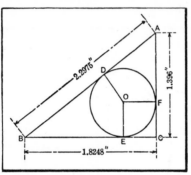

to be 43 degrees, 43 minutes and angle ($A - x$), to be 35 degrees, 10 minutes. By subtracting these two values, angle x will be found to equal 8 degrees, 33 minutes.

Example 5: — In tool designing it frequently becomes necessary to determine the length of a tangent to two circles. In Fig. 4, R = radius of large circle = $\frac{13}{16}$ inch; r =

Fig. 5. To find Radius of Circle Inscribed in Triangle

radius of small circle = $\frac{3}{8}$ inch; W = center distance between circles = $1\frac{11}{16}$ inches.

With the values given it is required to find the following: E = length of tangent; B = length of horizontal line from point of tangency on large circle to the vertical center line; and C = length of horizontal line from point of tangency on small circle to the vertical center line.

$$\sin a = \frac{R - r}{W} = \frac{\frac{13}{16} - \frac{3}{8}}{1\frac{11}{16}} = 0.25925$$

Angle $a = 15°\ 1'$ nearly

$$E = W \cos a = 1\frac{11}{16} \times 0.9658 = 1.63 \text{ inches}$$

$$B = R \sin a \qquad \text{and} \qquad C = r \sin a$$

Example 6: — In a right triangle having the dimensions shown in Fig. 5, a circle is inscribed. Find the radius of the circle.

In the illustration, $BD = BE$ and $AD = AF$, because "tangents drawn to a circle from the same point are equal." $EC =$

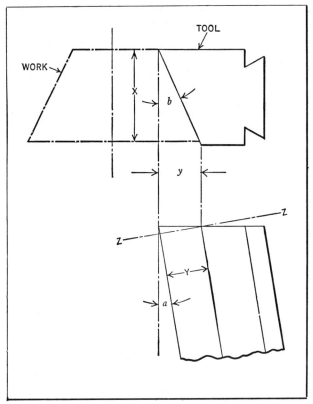

Fig. 6. The Problem is to Determine Angle of Forming Tool
in Plane *Z-Z*

CF, and EC = radius OF. Then let R = radius of inscribed circle. $AC - R = AD$ and $BC - R = DB$. Adding,

$$AC + BC - 2\,R = AD + DB$$
$$AD + DB = AB$$
hence, $$AC + BC - AB = 2\,R$$

Stated as a rule, "The diameter of a circle inscribed in a right triangle is equal to the difference between the hypotenuse and the sum of the other sides." Substituting the given dimensions, we have $1.396 + 1.8248 - 2.2975 = 0.9233 = 2R$, and $R = 0.4616$.

Example 7: — A part is to be machined to an angle b of 30 degrees (Fig. 6), by using a vertical forming tool having a clearance angle a of 10 degrees. Calculate the angle of the forming tool as measured in a plane $Z-Z$ which is perpendicular to the front or clearance surface of the tool.

Assume that B represents the angle in plane $Z-Z$.

$$\tan B = \frac{Y}{X} \quad \text{and} \quad Y = y \times \cos a \tag{1}$$

Also,

$$y = X \times \tan b \quad \text{and} \quad X = \frac{y}{\tan b} \tag{2}$$

Now substituting the values of Y and X in Equation (1), we have:

$$\tan B = \frac{y \times \cos a}{\dfrac{y}{\tan b}}$$

Clearing this equation of fractions,

$$\tan B = \cos a \times \tan b$$

In this example, $\tan B = 0.98481 \times 0.57735 = 0.56858$; hence $B = 29° \ 37'$ nearly.

Example 8: — A method of checking the diameter at the small end of a taper plug gage is shown by Fig. 7. The gage is first mounted on a sine-bar so that the top of the gage is parallel with the surface plate. A disk of known radius r is then placed in the corner formed by the end of the plug gage and the top side of the sine-bar. Now by determining the difference X in height between the top of the gage and the top edge of the disk, the accuracy of the diameter B can be checked readily. Derive formulas for determining dimension X.

The known dimensions are:

e = angle of taper;
r = radius of disk; and
B = required diameter at end of plug gage.

$$g = 90 \text{ degrees} - \tfrac{1}{2}e \quad \text{and} \quad k = \tfrac{1}{2}g$$

By trigonometry,

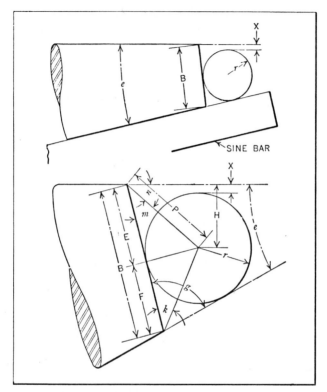

Fig. 7. The Problem is to Determine Height X in Order to Check Diameter B of Taper Plug

$$F = \frac{r}{\tan k}; \quad E = B - F; \quad \text{and} \quad \tan m = \frac{r}{E}$$

Also

$$P = \frac{r}{\sin m}; \quad n = g - m; \quad \text{and} \quad H = P \sin n$$

Therefore, $X = H - r$ or $r - H$, depending on whether or not the top edge of the disk is above or below the top of the plug gage. In the illustration the top of the disk is below the top surface of the plug gage so that it is evident that $X = H - r$.

To illustrate the application of these formulas, assume that $e = 6$ degrees, $r = 1$ inch and $B = 2.400$ inches. The dimension X is then found as follows:

$$g = 90 - \frac{6}{2} = 87°; \quad \text{and} \quad k = 43° 30'$$

By trigonometry,

$$F = \frac{1}{0.94896} = 1.0538''; \quad E = 2.400 - 1.0538 = 1.3462 \text{ inches}$$

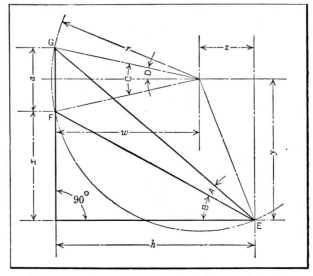

Fig. 8. Find Dimension x and Angle B, having a, h, and Angle A

$$\tan m = \frac{1}{1.3462} = 0.74283 \quad \text{and} \quad m = 36° 36' 22''$$

$$P = \frac{1}{0.59631} = 1.6769''; \quad n = 87° - 36° 36' 22'' = 50° 23' 38''$$

and

$$H = 1.6769 \times 0.77044 = 1.2920 \text{ inches}$$

Therefore,

$$X = H - r = 1.2920 - 1 = 0.2920 \text{ inch}$$

The disk in this case is below the top surface of the plug gage, hence the formula $X = H - r$ was applied.

Example 9: — In Fig. 8, $a = 1\frac{1}{4}$ inches, $h = 4$ inches, and angle $A = 12$ degrees. Find dimension x and angle B.

Draw an arc through points E, F, and G, as shown, with r as a radius. According to a well-known theorem of geometry, which is given on page 281 of MACHINERY'S HANDBOOK, if an angle at the circumference of a circle, between two chords, is subtended by the same arc as the angle at the center, between two radii, then the angle at the circumference is equal to one-half the angle at the center. This being true, angle C is twice

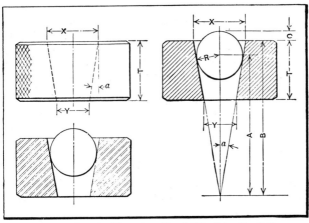

Fig. 9. Checking Dimensions X and Y by Using One Ball of Given Size

the magnitude of angle A, and angle D = angle A = 12 degrees. It will now be readily observed that

$$r = \frac{a}{2 \sin D} = \frac{1.25}{2 \times 0.20791} = 3.0061$$

$$w = \frac{a}{2} \cot D = 0.625 \times 4.7046 = 2.9404$$

and

$$z = h - w = 4 - 2.9404 = 1.0596$$

Now

$$y = \sqrt{r^2 - z^2} = \sqrt{7.9138505} = 2.8131$$

and

$$x = y - \frac{a}{2} = 2.8131 - 0.625 = 2.1881 \text{ inches}$$

Finally,

$$\tan B = \frac{x}{h} = \frac{2.1881}{4} = 0.54703$$

and

$$B = 28 \text{ degrees, } 40 \text{ minutes, } 47 \text{ seconds}$$

Example 10: — A steel ball is placed inside of a taper gage as shown in Fig. 9. If the angle of the taper, length of taper, radius of ball and its position in the gage are known, how can the end diameters X and Y of the gage be determined by measuring dimension C?

The ball should be of such size as to project above the face of the gage. This, however, is not necessary, although preferable, as it permits the required measurements to be more readily obtained. After measuring the distance C, the calculation of dimension X is as follows: First obtain dimension A, which equals R multiplied by csc a. Then adding R to A and subtracting C we obtain dimension B. Dimension X may then be obtained by multiplying $2 B$ by the tangent of angle a. The formulas for X and Y can therefore be written as follows:

$$X = 2\,(R \csc a + R - C) \tan a = 2\,R \sec a + 2 \tan a\,(R - C)$$
$$Y = X - (2\,T \tan a)$$

If in Fig. 9 angle $a = 9$ degrees, $T = 1.250$ inches, $C = 0.250$ inch and $R = 0.500$ inch, what are the dimensions X and Y? Applying the formula,

$$X = 2 \times 0.500 \times 1.0125 + 2 \times 0.15838\,(0.500 - 0.250)$$

Solving this equation, $X = 1.0917$ inches. Then

$$Y = 1.0917 - (2.500 \times 0.15838) = 0.6957$$

Example 11: — In designing a motion of the type shown in Fig. 10, it is essential, usually, to have link E swing equally above and below the center line MM. A mathematical solution of this problem follows. In the illustration, G represents the machine frame; F, a lever shown in the extreme positions; E, a link; and D, a slide. The distances A and B are fixed and the problem is to obtain $A + X$, or the required length of the lever. In the right triangle:

$$A + X = \sqrt{(A - X)^2 + \left(\frac{B}{2}\right)^2}$$

Squaring, we have:

$$A^2 + 2AX + X^2 = A^2 - 2AX + X^2 + \frac{B^2}{4}$$

$$4AX = \frac{B^2}{4}$$

$$X = \frac{B^2}{16A}$$

$$A + X = A + \frac{B^2}{16A} = \text{length of lever}$$

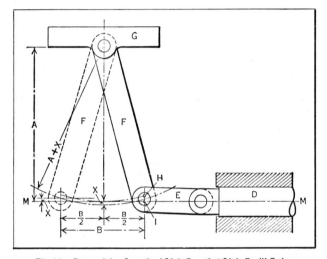

Fig. 10. Determining Length of Link *F* so that Link *E* will Swing Equally Above and Below the Center Line

To illustrate the application of this formula, assume that the length of a lever is required when the distance $A = 10$ inches and the stroke B of the slide is 4 inches.

Length of lever $= A + \dfrac{B^2}{16A} = 10 + \dfrac{16}{16 \times 10} = 10.100$ inches

Thus it is evident that the pin in the lower end of the lever will be 0.100 inch below the center line $M-M$ when half the stroke has been made, and at each end of the stroke the pin will be 0.100 inch above this center line.

Example 12: — The spherical hubs of bevel gears are checked by measuring the distance x (Fig. 11) over a ball or plug placed against a plug gage which fits into the bore. Determine this distance x.

First find H by means of the formula for circular segments on Handbook page 152.

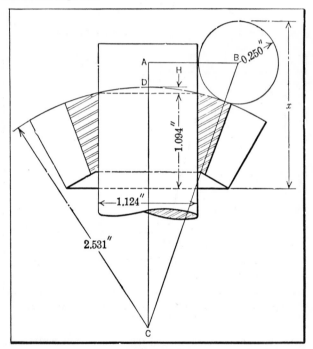

Fig. 11.　Method of Checking the Spherical Hub of a Bevel Gear with Plug Gages

$$H = 2.531 - \tfrac{1}{2}\sqrt{4 \times 2.531^2 - 1.124^2} = 0.0632 \text{ inch}$$

$$AB = \frac{1.124}{2} + 0.25 = 0.812 \text{ inch}$$

$$BC = 2.531 + 0.25 = 2.781 \text{ inch}$$

Applying one of the formulas for right triangles, on Handbook page 174,

$AC = \sqrt{2.781^2 - 0.812^2} = 2.6599$ inches

$AD = AC - DC = 2.6599 - 2.531 = 0.1289$ inch

$x = 1.094 + 0.0632 + 0.1289 + 0.25 = 1.536$ inches

Example 13: — The accuracy of a gage is to be checked by placing a ball or plug between the gage jaws and measuring to

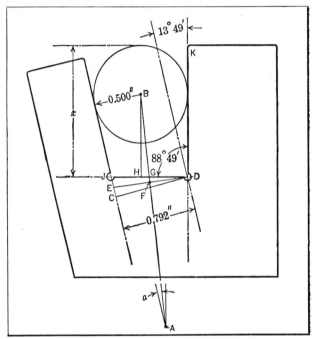

Fig. 12.   Finding Dimension *x* to Check Accuracy of Gage

the top of the ball or plug as shown by Fig. 12.   Dimension *x* is required and the known dimensions and angles are shown by the illustration.

One-half of the included angle between the gage jaws equals one-half of 13° 49′ or 6° 54½′ and the latter equals angle *a*

$$AB = \frac{0.500}{\sin 6° 54\frac{1}{2}'} = 4.1569 \text{ inches}$$

DE is perpendicular to *AB* and angle *CDE* = angle *a*; hence,

$$DE = \frac{CD}{\cos 6^\circ \, 54\frac{1}{2}'} = \frac{0.792}{\cos 6^\circ \, 54\frac{1}{2}'} = 0.79779 \text{ inch}$$

$$AF = \frac{DE}{2} \times \cot 6^\circ \, 54\frac{1}{2}' = 3.2923 \text{ inches}$$

Angle $CDK = 90^\circ + 13^\circ \, 49' = 103^\circ \, 49'$

Angle $CDJ = 103^\circ \, 49' - 88^\circ \, 49' = 15^\circ$

Angle $EDJ = 15^\circ - 6^\circ \, 54\frac{1}{2}' = 8^\circ \, 5\frac{1}{2}'$

$$GF = \frac{DE}{2} \times \tan 8^\circ \, 5\frac{1}{2}' = 0.056711 \text{ inch}$$

Angle $HBG =$ angle $EDJ = 8^\circ \, 5\frac{1}{2}'$

$BG = AB - (GF + AF) = 0.807889$ inch

$BH = BG \times \cos 8^\circ \, 5\frac{1}{2}' = 0.79984$ inch

$\quad x = BH + 0.500 = 1.2998$ inches

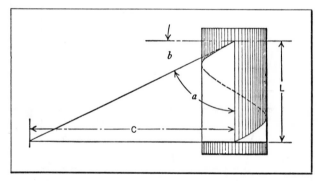

Fig. 13. Helix Represented by a Triangular Piece of Paper Wound Upon a Cylinder

If surface JD is parallel to the bottom surface of gage, the distance between these surfaces might be added to x in order to use a height gage from a surface plate.

Helix Angles of Screw Threads, Hobs, and Helical Gears. — The terms "helical" and "spiral" often are used interchangeably in drafting-rooms and shops, although the two curves are entirely different. As the illustration on Handbook page 287 shows, every point on a helix is equidistant from the axis, and the curve advances at a uniform rate around a cylindrical area. The helical and spiral springs illustrated on Handbook page 499 show the difference between a helix and a spiral. A spiral may be defined mathematically as a curve having a constantly increasing radius of curvature.

If a piece of paper is cut in the form of a right triangle and wrapped around a cylinder, as indicated by the diagram (Fig. 13), the hypotenuse will form a helix. The curvature of a screw thread represents a helix. From the properties of a right triangle, simple formulas can be derived for determining helix angles. Thus, if the circumference of a part is divided by the lead or distance that the helix advances axially in one turn, the quotient equals the tangent of the helix angle as measured from

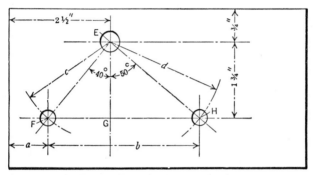

Fig. 14.   Find Dimensions a, b, c, and d

the axis. The angles of helical curves usually are measured from the axis but not invariably. The helix angle of a helical or "spiral" gear is measured from the axis but the helix angle of a screw thread is measured from a plane perpendicular to the axis. In case of a helical gear, the angle is a (Fig. 13), whereas, for a screw thread, the angle is b; hence, for helical gears tan a of helix angle $= \dfrac{C}{L}$; for screw threads, tan b of helix angle $= \dfrac{L}{C}$. The helix angle of a hob such as is used for gear cutting, also is measured as indicated at b and often is known as the "end angle" because it is measured from the plane of the end surface of the hob. In calculating helix angles of helical gears, screw threads, and hobs, the pitch circumference is used.

Example 14: — If the pitch diameter of a helical gear = 3.818 inches and the lead = 12 inches, what is the helix angle?

tan helix angle $= \dfrac{3.818 \times 3.1416}{12} = 1$ very nearly; hence the angle $= 45$ degrees.

PRACTICE EXERCISES FOR SECTION 9

For answers to all practice exercise problems or questions
see Section 20

1. The No. 4 Morse taper is 0.6233 inch per foot: calculate the
   included angle.

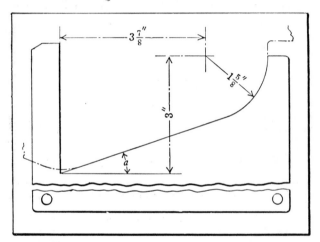

Fig. 15.   To find Angle a having the Dimensions Given

2. Briggs standard pipe threads have a taper of $\frac{3}{4}$ inch per
   foot. What is the angle on each side of the center line?
3. To what dimension should the dividers be set to space 8
   holes evenly on a circle of 6 inches diameter?
4. Explain the derivation of the formula

$$W = \sin \left(\dfrac{\dfrac{360^\circ}{N} - 2\,a}{2} \right) \times B$$

For notation, see Example 2 of Section 9 and the
diagram Fig. 1.

5. The top of a male dovetail is 4 inches wide. If the angle is 55 degrees and the depth is $\frac{5}{8}$ inch, what is the width at the bottom of the dovetail?

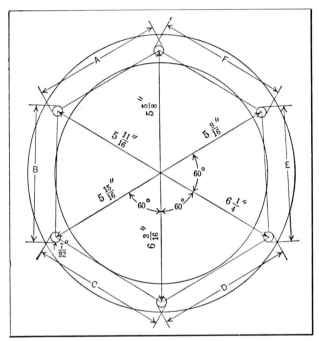

Fig. 16. To Find the Chordal Distances of Irregularly Spaced Holes Drilled in a Taximeter Drive Ring

6. Angles may be laid out accurately by describing an arc with a radius of given length and then determining the length of a chord of this arc. In laying out an angle of 25 degrees, 20 minutes, using a radius of 8 inches, what should be the length of the chord opposite the angle named?

7. How large a square may be milled on the end of a $2\frac{1}{2}$-inch bar of round stock?

8. A guy wire from a smoke stack is 120 feet long. How high

is the stack if the wire is attached 10 feet from the top and makes an angle of 57 degrees with the stack?

9. In laying out a master jig plate, it is required that holes *F* and *H*, Fig. 14, shall be on a straight line which is $1\frac{3}{4}$ inch distant from hole *E*. The holes must also be on lines making, respectively, 40- and 50-degree angles with line *EG*, drawn at right angles to the sides of the jig plate through *E*, as shown in the engraving. Find the dimensions *a*, *b*, *c* and *d*.

10. In Fig. 15 is shown a template for locating a pump body on a milling fixture, the inside contour of the template corresponding with the contour of the pump flange. Find the angle *a* from the values given.

11. Find the chordal distances as measured over plugs placed in holes located at different radii in the taximeter drive ring shown in Fig. 16. All holes are $\frac{7}{32}$ inch diameter; the angle between the center line of each pair of holes is 60 degrees.

12. An Acme screw thread has an outside diameter of $1\frac{1}{4}$ inches and has 6 threads per inch. Find the helix angle using the pitch diameter as a base. Find, also, the helix angle if a double thread is cut on the screw.

13. What is the lead of the flutes in a $\frac{7}{8}$-inch drill if the helix angle, measured from the center line of the drill, is 27° 30′?

14. A 4-inch diameter milling cutter has a lead of 68.57 inches. What is the helix angle measured from the axis?

SECTION 10

SOLUTION OF OBLIQUE TRIANGLES

Handbook Page 176

In solving problems for dimensions or angles it is often convenient to work with oblique triangles. In an oblique triangle none of the angles are right angles. One of the angles may be over 90 degrees or each of the three angles may be less than 90 degrees. Any oblique triangle may be solved by constructing perpendiculars to the sides from appropriate vertices, thus forming right triangles. The methods, previously explained, for solving right triangles, will then solve the oblique triangles. The objection to this method of solving oblique triangles is that it is a long, tedious process.

Two of the examples in the handbook on page 177 which are solved by the formulas for oblique triangles, will be solved by the right-angle triangle method. These have been solved to show that all oblique triangles can be thus solved and to give an opportunity to compare the two methods. All oblique triangles come under four classes:

(1) Given one side and two angles.
(2) Given two sides and the included angle.
(3) Given two sides and the angle opposite one of them.
(4) Given the three sides.

Example 1: — Solve the first example on Handbook page 177 by the right-angle triangle method. Referring to the accompanying Fig. 1.

$$\text{Angle } C = 180° - (62° + 80°) = 38°$$

Draw a line DC perpendicular to AB.

In the right triangle BDC, $\dfrac{DC}{BC} = \sin 62°$

$\dfrac{DC}{5} = 0.88295$; $DC = 5 \times 0.88295 = 4.41475$

Angle $BCD = 90° - 62° = 28°$; $DCA = 38° - 28° = 10°$

$\dfrac{BD}{5} = \cos 62°$; $BD = 5 \times 0.46947 = 2.34735$

In triangle ADC, $\dfrac{AC}{DC} = \sec 10°$

$AC = 4.41475 \times 1.0154 = 4.4827$

$\dfrac{AD}{4.41475} = \tan 10°$; $AD = 4.41475 \times 0.17633 = 0.7785$

and $AB = AD + DB = 0.7785 + 2.34735 = 3.1258$

$C = 38°$; $b = 4.4828$; $c = 3.1258$

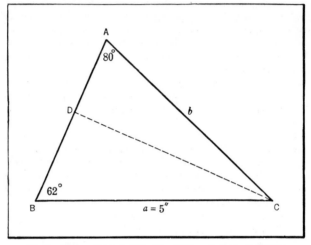

Fig. 1.   Oblique Triangle Solved by Right-angle Triangle Method

Example 2: — Apply the right-angle triangle method to the solution of the second example on Handbook page 177.

Referring to Fig. 2, draw a line BD perpendicular to CA.

In the right triangle BDC, $\dfrac{BD}{9} = \sin 35°$

$BD = 9 \times 0.57358 = 5.16222$

$\dfrac{CD}{9} = \cos 35°$; $CD = 9 \times 0.81915 = 7.37235$

$DA = 8 - 7.37235 = 0.62765$

In the right triangle BDA, $\dfrac{BD}{DA} = \dfrac{5.16222}{0.62765} = \tan A$

$\tan A = 8.2246$ and $A = 83° 4'$

$B = 180° - (83° 4' + 35°) = 61° 56'$

$\dfrac{BA}{BD} = \dfrac{BA}{5.1622} = \text{cosec } 83° 4'; \ BA = 5.1622 \times 1.0074 = 5.2004$

$A = 83° 4'; \ B = 61° 56'; \ C = 35°$

$a = 9; \ b = 8; \ c = 5.2004$

Use of Formulas for Oblique Triangles. — Oblique triangles are not encountered as frequently as right triangles and therefore the methods of solving the latter may be fresh in the memory while methods for solving the former may be forgotten. All

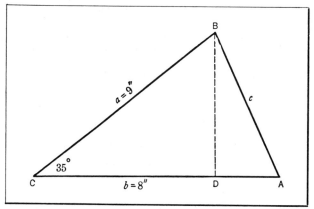

Fig. 2.   Another Example of the Right-angle Triangle Solution of an Oblique Triangle

the formulas involved in the solution of the four cases of oblique triangles are derived from: (1) the law of sines; (2) the law of cosines and (3) the sum of the angles of a triangle equal 180°.

The law of sines is that in any triangle, the length of the sides are proportional to the sines of the opposite angles. (See diagrams on Handbook page 176.)

$\dfrac{a}{\sin A} = \dfrac{b}{\sin B} = \dfrac{c}{\sin C}$ (1). Solving this equation we get:

$\dfrac{a}{\sin A} = \dfrac{b}{\sin B}$ then $a \times \sin B = b \times \sin A$ and

$a = \dfrac{b \times \sin A}{\sin B}; \quad \sin B = \dfrac{b \times \sin A}{a}$

$$b = \frac{a \times \sin B}{\sin A} \; ; \quad \sin A = \frac{a \times \sin B}{b}$$

In like manner, $\dfrac{a}{\sin A} = \dfrac{c}{\sin C}$ and

$$a \times \sin C = c \times \sin A; \quad \text{hence} \quad \sin A = \frac{a \times \sin C}{c}$$

and $\dfrac{b}{\sin B} = \dfrac{c}{\sin C}$ or $b \times \sin C = c \times \sin B.$

Thus twelve formulas may be derived. As a general rule only formula (I) is remembered and special formulas are derived from it as required.

The law of cosines states that in any triangle the square of any side equals the sum of the squares of the other two sides minus twice their product multiplied by the cosine of the angle between them. These relations are stated as formulas thus:

(1) $a^2 = b^2 + c^2 - 2\,bc \times \cos A$ or $a = \sqrt{b^2 + c^2 - 2\,bc \times \cos A}$

(2) $b^2 = a^2 + c^2 - 2\,ac \times \cos B$ or $b = \sqrt{a^2 + c^2 - 2\,ac \times \cos B}$

(3) $c^2 = a^2 + b^2 - 2\,ab \times \cos C$ or $c = \sqrt{a^2 + b^2 - 2\,ab \times \cos C}$

Solving (I), $a^2 = b^2 + c^2 - 2\,bc \times \cos A$ for $\cos A$,

$2\,bc \times \cos A = b^2 + c^2 - a^2$ (transposing)

$\cos A = \dfrac{b^2 + c^2 - a^2}{2\,bc}$

In like manner formulas for $\cos B$ and $\cos C$ may be found.

Example 3: — A problem quite often encountered in lay-out work is illustrated in Fig. 3. It is required to find the dimensions *x* and *y* between the holes, these dimensions being measured from the intersection of the perpendicular line with the center line of the two lower holes. The three center-to-center distances are the only known values.

The method that might first suggest itself is to find the angle *A* (or *B*) by some such formula as

$$\cos A = \frac{b^2 + c^2 - a^2}{2\,bc} \tag{1}$$

and then solve the right triangle for *y* by the formula

$$y = b \cos A \tag{2}$$

Formulas (I) and (2) can be combined as follows:

$$y = \frac{b^2 + c^2 - a^2}{2\,c} \tag{3}$$

The value of x can be determined in a similar manner.

A second solution of this problem involves the following geometrical proposition: In any oblique triangle where the three sides are known, the ratio of the length of the base to the sum of the other two sides equals the ratio of the difference between

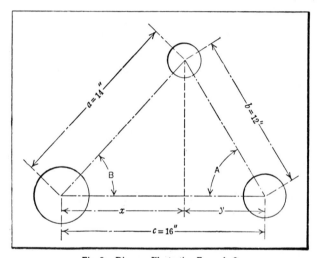

Fig. 3. Diagram Illustrating Example 3

the length of the two sides to the difference between the lengths x and y. Therefore, if $a = 14$, $b = 12$ and $c = 16$ inches, then

$$c : (a + b) = (a - b) : (x - y)$$
$$16 : 26 = 2 : (x - y)$$
$$(x - y) = \frac{26 \times 2}{16} = 3\tfrac{1}{4} \text{ inches}$$
$$x = \frac{(x + y) + (x - y)}{2} = \frac{16 + 3\tfrac{1}{4}}{2} = 9.625 \text{ inches}$$
$$y = \frac{(x + y) - (x - y)}{2} = \frac{16 - 3\tfrac{1}{4}}{2} = 6.375 \text{ inches}$$

When Angles have Negative Values. — In the solution of oblique triangles having one angle larger than 90 degrees, it is sometimes necessary to use angles whose functions are negative. Review Handbook pages 105 and 178. Notice that for angles between 90 degrees and 180 degrees the cosine, tangent, cotangent and secant are negative.

Example 4: — Referring to Fig. 4, two sides and the angle between them are known. Find angles A and B. (See Handbook page 176.)

$$\tan A = \frac{4 \times \sin 20°}{3 - 4 \times \cos 20°} = \frac{4 \times 0.34202}{3 - 4 \times 0.93969} = \frac{1.36808}{3 - 3.75876}$$

It will be seen that in the denominator of the fraction above, the number to be subtracted from 3 is greater than 3; the num-

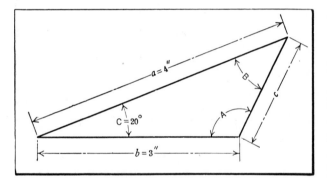

Fig. 4. Finding Angles A and B from the Dimension Given

bers are therefore reversed, 3 being subtracted from 3.75876, the remainder then being negative. Hence:

$$\tan A = \frac{1.36808}{3 - 3.75876} = \frac{1.36808}{-0.75876} = -1.80305$$

The final result is negative because a positive number (1.36808) is divided by a negative number (−0.75876). The tangents of angles greater than 90 degrees and smaller than 180 degrees are negative. To find an angle whose tangent is negative, find in this case the value nearest to 1.80305 in the columns of tangents in the Handbook tables. It will be seen that the nearest value is 1 8028, which is the tangent of 60° 59′. As the tangent

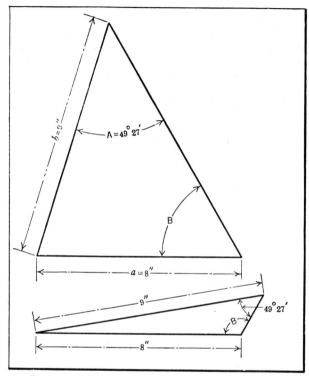

Fig. 5. Diagrams Showing Two Possible Solutions of the Same Problem which is to Find Angle B

here is negative, angle A, however, is not 60° 59', but equals 180° − 60° 59' = 119° 1'. Now angle B is found by the formula

$$B = 180° - (A + C) = 180° - (119° 1' + 20°) =$$
$$180° - 139° 1' = 40° 59'$$

When Either of Two Triangles Conform to the Given Dimensions. — When two sides and the angle opposite one of the given sides are known, *if the side opposite the given angle is shorter than the other given side,* two triangles can be drawn as shown by Fig. 5 which have sides of the required length and the required angle opposite one of the sides. The lengths of the two known

sides of each triangle are 8 and 9 inches, and the angle opposite
the 8-inch side is 49° 27′ in each case; but it will be seen that
the angle B of the lower triangle is very much larger than the
corresponding angle of the upper triangle, and there is a great
difference in the areas. When two sides and one of the opposite
angles are given, the problem is capable of two solutions when
(and only when) the side opposite the given angle is shorter
than the other given side. When the triangle to be calculated
is drawn to scale, it is possible to determine from the shape of
the triangle which of the two solutions applies.

Example 5: — Find angle B, Fig. 5, from the formula, $\sin B = \dfrac{b \times \sin A}{a}$ where $b = 9$ inches; $A = 49$ degrees, 27 minutes; a is side opposite angle $A = 8$ inches.

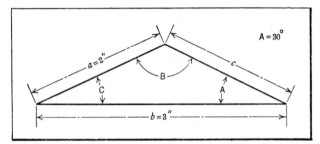

Fig. 6. Another Example Which has Two Possible Solutions

$$\sin B = \frac{9 \times 0.75984}{8} = 0.85482 = \sin 58° \ 44′ \ 24″ \text{ or } \sin 121°$$

15′ 36″. The practical requirements of the problem doubtless
will indicate which of the two triangles shown in Fig. 5 is the
right one.

Example 6: — In Fig. 6, $a = 2$ inches, $b = 3$ inches and $A = 30$
degrees. Find B.

$$\sin B = \frac{b \times \sin A}{a} = \frac{3 \times \sin 30°}{2} = 0.75000$$

We find from the tables that sine 0.75000 is the sine of 48° 35′.
From Fig. 6 it is apparent, however, that B is greater than 90
degrees, and as 0.75000 is the sine not only of 48° 35′, but also
of 180° − 48° 35′ = 131° 25′, angle B in this case equals 131° 25′.

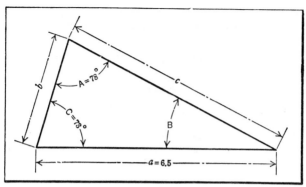

Fig. 7. Example for Practice Exercise No. 2

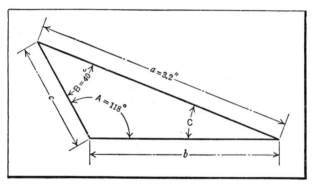

Fig. 8. Example for Practice Exercise No. 3

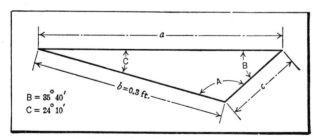

Fig. 9. Example for Practice Exercise No. 4

This example illustrates how the practical requirements of the problem, indicate which of two angles is correct.

PRACTICE EXERCISES FOR SECTION 10

For answers to all practice exercise problems or questions
see Section 20

1. Three holes in a jig are located as follows:
   Hole No. 1 is 3.375 from hole No. 2 and 5.625 from hole No. 3; the distance between No. 2 and No. 3 is 6.250. What three angles between the center lines are thus formed?

2. In Fig. 7 is shown a triangle one side of which is 6.5 feet, and the two angles A and C are 78 and 73 degrees, respectively. Find angle B, sides b and c, and the area.

3. In Fig. 8, side a equals 3.2 inches, angle A, 118 degrees, and angle B 40 degrees. Find angle C, sides b and c, and the area.

4. In Fig. 9, side $b = 0.3$ foot, angle $B = 35° 40'$, and angle $C = 24° 10'$. Find angle A, sides a and c, and the area.

5. Give two general rules for finding the areas of triangles.

SECTION 11

LOGARITHMS OF TRIGONOMETRICAL FUNCTIONS

Handbook Pages 224 to 269

Logarithms of trigonometrical functions are used for the same reason that common logarithms are used — they save time and simplify the work. As explained on Handbook page 224, the form of the characteristic of the logarithms of trigonometrical functions has been changed to avoid negative characteristics by adding 10. The −10 is omitted in these tables. Thus the log sine 8° 20′ is given as 9.16116 but the actual value is 9.16116 − 10 or 1.16116. These values are used in solving formulas exactly as explained in Section 5.

The application of logarithms to the solution of triangles is recommended especially when dealing with problems involving four or five place decimals, or whenever the use of logarithms will avoid tedious multiplications or divisions.

In Section 8 are given simple formulas relating to the different trigonometrical functions. One of these formulas shows that the sine of an angle is merely the ratio of the side opposite the angle to the hypotenuse; another shows that the cosine is the ratio of the side adjacent to the hypotenuse. Now, by transposing these simple expressions, we find, for example, that

$$\text{Side Opposite} = \text{Hypotenuse} \times \text{Sine}$$
$$\text{Side Adjacent} = \text{Hypotenuse} \times \text{Cosine}$$

On the accompanying diagram, Fig. 1, Hyp. represents hypotenuse; SO represents "side opposite"; and, SA, "side adjacent." If $SO = \text{Hyp} \times \text{sine } a$, then $\log SO = \log \text{Hyp} + \log \text{sine } a$, because (as explained on Handbook page 122) multiplication by logarithms involves adding the logarithms of the numbers to be multiplied, to obtain the logarithm of the product. In this manner all of the ten rules which follow have been developed. These general rules are for application to various problems in trigonometry, assuming that logarithms can be used to advantage.

1. $\log SO = \log \text{Hyp} + \log \sin a$
2. $\log SO = \log SA + \log \tan a$
3. $\log SA = \log \text{Hyp} + \log \cos a$
4. $\log SA = \log SO + \log \cot a$
5. $\log \text{Hyp} = \log SO - \log \sin a$
6. $\log \text{Hyp} = \log SA - \log \cos a$
7. $\log \sin a = \log SO - \log \text{Hyp}$
8. $\log \cos a = \log SA - \log \text{Hyp}$
9. $\log \tan a = \log SO - \log SA$
10. $\log \cot a = \log SA - \log SO$

Example 1: — Referring to Fig. 1, if $SO = 8.023$ inches and $SA = 15.781$ inches, what is the value of angle a?

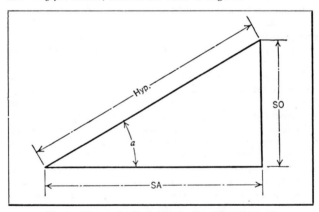

Fig. 1.   Diagram for Illustrating the Use of Logarithms of Trigonometrical Functions

The first rule for determining angle a is No. 7 in the preceding list of ten rules, but this rule cannot be used because the hypotenuse is not given; hence, Rule 9 is applied.

$$\log \tan a = \log SO - \log SA$$
$$\log \tan a = \log 8.023 - \log 15.781$$

By referring to the table of common logarithms beginning on Handbook page 126, we find that

$$\log 8.023 = 0.90434$$
$$\log 15.781 = 1.19814$$

If the form of the logarithm 0.90434 is changed by adding
10 to its value to avoid dealing with a negative characteristic,
then,

$$\begin{array}{rl} \log\ 8.023 = & 10.90434 \\ \log\ 15.781 = & 1.19814 \\ \hline \log\ \tan a = & 9.70620 \end{array}$$

The angle equivalent to 9.70620 is 26° 57′. (See Handbook
page 251.)

If in the foregoing example the logarithm of 1.19814 is given
a negative value, then the numbers 0.90434 and $\bar{2}$.80186 can be
added instead of subtracted; thus

$$\begin{array}{rl} \log\ 8.023 = & 0.90434 \\ \log\ 15.781 = & \bar{2}.80186 \\ \hline & \bar{1}.70620 \end{array}$$

As the logarithms in the tables beginning page 225 have had
10 added to their values

$$\bar{1}.70620 + 10 = 9.70620$$

(See on page 121, "Avoiding Use of Negative Characteristics".)

Example 2:—Using the formula $\sin B = \dfrac{b \times \sin A}{a}$, find
angle B if $a = 7.194''$; $b = 6.782''$ and angle $A = 44°\ 29'$.

$$\begin{array}{rl} \log 6.782 = & 0.83136 \\ \log \sin 44°\ 29' = & 9.84553 \\ \hline \log\ \text{numerator} = & 10.67689 \\ \log 7.194 = & 0.85697 \\ \hline \log \sin B = & 9.81992 \end{array}$$

In this example the log 9.84553 which is given in the Handbook table for sin 44° 29′ is used without change. Since division
is required, the log of the divisor 7.194 is subtracted from the log
of the dividend or numerator in this case, and the difference
= 9.81992 = log sin B. The table shows that the corresponding
angle is 41° 21′. The actual value of log sin 44° 29′ is 9.84553 −
10; the log of numerator = 10.67689 − 10 and log sin B =
9.81992 − 10, but in this example the omission of −10 does not
affect the result.

If, in connection with Example 2, the actual values of the

logarithms are used, as illustrated in the Handbook, page 224, we have

$$\begin{aligned}
\log 6.782 &= 0.83136 \\
\log \sin 44° \ 29' &= \bar{1}.84553 \\
- \log 7.194 &= \bar{1}.14303 \\
\hline
\log \sin B &= \bar{1}.81992
\end{aligned}$$

In this case the three logs are merely added and in finding the angle corresponding to log sin B, the value 9.81992 is used because $\bar{1}.81992 + 10 = 9.81992 = \log$ as given in the table of logarithms of trigonometrical functions.

Example 3: — Using the formula, log side adjacent = log hypotenuse + log cos A, find the side adjacent when the hypotenuse = 65.4 inches and $A = 52° \ 12'$.

In this example a dimension or the length of the side is required so that the logarithm corresponding to this dimension must be the actual logarithm of the number since the regular logarithm tables (beginning on Handbook page 126) must be used to find this number.

$$\begin{aligned}
\log 65.4 &= 1.81558 \\
\log \cos 52° \ 12' &= \bar{1}.78739 \\
\hline
\log \text{ side adjacent} &= 1.60297
\end{aligned}$$

Assume now that the log cos 52° 12′ is used as given in the table (see Handbook page 262 and note that for angles from 45° to 90° the degrees are in the lower right-hand corner, the functions are at the bottom of the table, and the minutes on the right-hand side should be read up from the bottom.)

$$\begin{aligned}
\log 65.4 &= \ \ 1.81558 \\
\log \cos 52° \ 12' &= \ \ 9.78739 \\
\hline
&\ 11.60297
\end{aligned}$$

Since the true value of log cos 52° 12′ is 9.78739 − 10, the log for the side adjacent = 11.60297 − 10 = 1.60297, the result being the same as obtained by the first method. The number corresponding to log 1.60297 is 40.084, (see Handbook page 132) which is the length of the adjacent side of the triangle.

In many problems involving trigonometry logarithms may be used. In order to illustrate the simplicity of logarithms as applied to trigonometry, find, for example, the sine of 26° 28′; this equals 0.44568; now find the log of 0.44568. This logarithm equals $\overline{1}.64902$. Check this result by finding the log sin 26° 28′ in the table on page 251. The sine given in this table is 9.64902 which actually is $9.64902 - 10 = \overline{1}.64902$.

Example 4: — Two sides *a* and *b* of a triangle are 9 and 17 inches long respectively. (See diagram Fig. 2.) The angle *C* included between them is 32 degrees. Find the angle opposite the side 9 inches long.

The formula by means of which this angle can be found is:

$$\tan A = \frac{a \times \sin C}{b - a \times \cos C} = \frac{9 \times \sin 32°}{17 - 9 \times \cos 32°}$$

As only multiplication and division can be carried out by means of ordinary logarithms, the subtraction in the denominator must be made independently of logarithms; but logarithms can be used for the multiplications and divisions required. The first

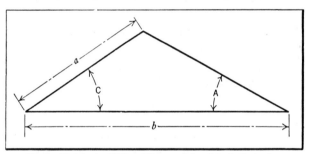

Fig. 2. Find Angle *A* When Sides *a* and *b* and Angle *C* are given

step will be to find the value of the denominator; we must then first find the product 9 × cos 32°.

$$\log 9 = 0.95424$$
$$\log \cos 32° = \overline{1}.92842$$
$$\overline{0.88266}$$

Hence $9 \times \cos 32° = 7.6323$, and $17 - 7.6323 = 9.3677$. Therefore.

$$\tan A = \frac{9 \times \sin 32°}{9.3677}$$

$$\log 9 = 0.95424$$
$$\log \sin 32° = \overline{1}.72421$$
$$-\log 9.3677 = \overline{1}.02837$$
$$\overline{\overline{}\overline{1}.70682}$$

Log tan A = $\bar{1}$.70682, or as given in the tables 9.70682; hence A = 26° 59'.

PRACTICE EXERCISES FOR SECTION 11

For answers to all practice exercise problems or questions
see Section 20

1. Find the log sin 42° 18'; 37° 46'.

2. Find the log cos 27° 15'; 48° 19'.

3. Find the log tan 41° 10'; 82° 52'.

4. What angles correspond to the following:
   log sin = 9.70547?; log sin = 9.85012?

5. Find corresponding angles if log cos = 9.94448; log sec = 10.03416.

6. Solve for c in the formula: $c = \dfrac{a \times \sin C}{\sin A}$ if $a = 9$, $C = 35°$; $A = 83° 4'$.

7. The spring pressure (P) in pounds for cone clutches is calculated from the formula:

$$P = \frac{H \times 63{,}025 \times \sin a}{f \times r \times n},$$

where H = horsepower; a = angle of cone from axis; f = coefficient of friction; r = radius of cone, and n = revolutions per minute. Find the spring pressure if $a = 12°$; $f = 0.25$; $r = 15.5$; $n = 2500$; $H = 60$.

8. Find side b in an oblique angle triangle, using the formula $b = \dfrac{a \times \sin B}{\sin A}$, where $a = 15$; $B = 70° 20'$ and A is angle opposite side $a = 58° 32'$.

SECTION 12

FIGURING TAPERS

Handbook Pages 1557 to 1561

The term "taper," as applied in shops and drafting-rooms, means the difference between the large and small dimensions where the increase in size is uniform. Since tapering parts generally are conical, taper means the difference between the large and small diameters. Taper is ordinarily expressed as a certain number of inches per foot; thus, $\frac{1}{2}''$ per ft.; $\frac{3}{4}''$ per ft.; etc. In certain kinds of work, taper is also expressed as a decimal part of an inch per inch, as: $0.050''$ per inch. The length of the work is always measured parallel to the center-line (axis) of the work, and never along the tapered surface.

Suppose that the diameter at one end of a tapering part is one inch, and the diameter at the other end, one and one-half inches, and that the length of the part is one foot. This piece, then, tapers one-half inch per foot, because the difference between the diameters at the ends is one-half inch. If the diameters at the ends of a part are $\frac{7}{16}$ inch and $\frac{1}{2}$ inch, and the length is one inch, this piece tapers $\frac{1}{16}$ inch per inch. The usual problems met with in regard to figuring tapers may be divided into seven classes. The rule to be used in each case may be found on Handbook page 1557.

Example 1: — The diameter at the large end of a part is $2\frac{5}{8}$ inches, the diameter at the small end, $2\frac{3}{16}$ inches, and the length of the work, 7 inches. Find the taper per foot.

Referring to the third rule on Handbook page 1557,

$$\text{Taper per foot} = \frac{2\frac{5}{8} - 2\frac{3}{16}}{7} \times 12 = \frac{3}{4} \text{ inch}$$

Example 2: — The diameter at the large end of a tapering part is $1\frac{5}{8}$ inches, the length is $3\frac{1}{2}$ inches, and the taper per foot is $\frac{3}{4}$ inch. The problem is to find the diameter at the small end.

Applying the fourth rule on Handbook page 1557,

$$\text{Diameter at small end} = 1\frac{5}{8} - \left(\frac{\frac{3}{4}}{12} \times 3\frac{1}{2} \right) = 1\frac{13}{32} \text{ inch}$$

87

Example 3: — What is the length of the taper if the two end diameters are 2.875 inches and 2.542 inches, the taper per foot being one inch?

Applying the sixth rule on Handbook page 1557,

$$\text{Distance between the two diameters} = \frac{2.875 - 2.542}{1} \times 12 = 4$$

inches nearly

Example 4: — If the length of the taper is 10 inches and the taper per foot is $\frac{3}{4}$ inch, what is the taper in the given length?

Applying the last rule on Handbook page 1557,

$$\text{Taper in given length} = \frac{\frac{3}{4}}{12} \times 10 = 0.625 \text{ inch}$$

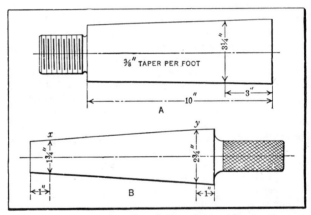

Fig. 1. Illustrations for Examples 6 and 7

Example 5: — The small diameter is 1.636 inches, the length of the work is 5 inches, and the taper per foot is $\frac{1}{4}$ inch; what is the large diameter?

Referring to the fifth rule on Handbook page 1557,

$$\text{Diameter at large end} = \left(\frac{\frac{1}{4}}{12} \times 5\right) + 1.636 = 1.740 \text{ inches}$$

Example 6: — Sketch *A*, Fig. 1, shows a part used as a clamp bolt. The diameter, $3\frac{1}{4}$ inches, is given 3 inches from the large end of the taper. The total length of the taper is 10 inches.

The taper is $\frac{3}{8}$ inch per foot. Find the diameters at the large and small ends of the taper.

First find the diameter of the large end using the fifth rule on Handbook page 1557.

$$\text{Diameter at large end} = \left(\frac{\frac{3}{8}}{12} \times 3\right) + 3\tfrac{1}{4} = 3\tfrac{11}{32} \text{ inches}$$

To find the diameter at the small end, use the fourth rule on the Handbook page mentioned.

$$\text{Diameter at small end} = 3\tfrac{11}{32} - \left(\frac{\frac{3}{8}}{12} \times 10\right) = 3\tfrac{1}{32} \text{ inches}$$

Example 7: — At B, Fig. 1, is shown a taper master gage intended for inspecting taper ring gages of various dimensions. The smallest diameter of the smallest ring gage is $1\tfrac{3}{4}$ inches, and the largest diameter of the largest ring gage is $2\tfrac{3}{4}$ inches. The taper per foot is $1\tfrac{1}{2}$ inches. It is required that the master gage extend one inch outside of the ring gages at both the small and the large ends, when these ring gages are tested. How long should the taper on the master gage be?

The sixth rule on Handbook page 1557 may be applied here.

$$\text{Distance between the two diameters} = \frac{2\tfrac{3}{4} - 1\tfrac{3}{4}}{1\tfrac{1}{2}} \times 12 = 8 \text{ inches}$$

$$\text{Total length of taper} = 8 + 2 = 10 \text{ inches}$$

Table for Converting Taper Per Foot to Degrees. — Some types of machines, such as milling machines, are graduated in degrees, making it necessary to convert the taper per foot to the corresponding angle in degrees. This is quickly done by means of the table, Handbook page 1558.

Example 8: — If a taper of $1\tfrac{1}{2}$ inches per foot is to be milled on a piece of work, at what angle must the machine table be set if the taper is measured from the axis of the work?

Referring to the Handbook table, the angle corresponding with a taper of $1\tfrac{1}{2}$ inches to the foot is $3° 34' 35''$ as measured from the center line.

Note that the taper per foot varies directly as *the tangent of one-half the included angle.* Two mistakes frequently made in figuring tapers are assuming that the taper per foot varies directly as the included angle or that it varies directly as the tangent of the included angle. In order to verify this point, refer to the table on Handbook page 1558 where it will be seen that the included angle for a taper of 4 inches per foot (18° 55′ 31″) is not twice the included angle for a taper of 2 inches per foot (9° 31′ 37″). Neither is the tangent of 18° 55′ 31″ (0.3428695) twice the tangent of 9° 31′ 37″ (0.1678261).

Tapers for Machine Tool Spindles. — The holes in machine tool spindles, for receiving tool shanks, arbors and centers, are tapering to insure a tight grip, accuracy of location, and also to facilitate removal of arbors, cutters, etc. The most common tapers are the Morse, the Brown & Sharpe, and the Jarno. The Morse has been very generally adopted for drilling machine spindles. Most engine lathe spindles also have the Morse taper, but some lathes have the Jarno or a modification of it, and others, a modified Morse taper which is longer than the standard. A standard milling machine spindle was adopted in 1927 by the milling machine manufacturers of the National Machine Tool Builders' Association. A comparatively steep taper of 3½ inches per foot was adopted in connection with this standard spindle to insure instant release of arbors. Prior to the adoption of the standard spindle, the Brown & Sharpe taper was used for practically all milling machines and this is also the taper for dividing-head spindles. There is considerable variation in grinding machine spindles. The Brown & Sharpe taper is the most common, but the Morse and the Jarno have also been used. Tapers of ⅝ inch per foot and ¾ inch per foot have also been used to some extent on miscellaneous classes of machines requiring a taper hole in the spindle.

PRACTICE EXERCISES FOR SECTION 12

For answers to all practice exercise problems or questions
see Section 20

1. What tapers, per foot, are used with the following tapers: (*a*) Morse taper; (*b*) Jarno taper; (*c*) milling machine spindle; (*d*) taper pin?

2. What is the taper per foot on a part if the included angle is 10° 30′; 55° 45′?

3. In setting up a taper gage like that shown on Handbook page 1559, what should be the center distance between 1.75″ and 2″ disks to check either the taper per foot or angle of a No. 4 Morse taper?

4. If it is required to check an angle of 14½°, using two disks in contact, and the smaller disk is 1″ diameter, what should be the diameter of the larger disk?

5. What should be the center distance, using disks of 2″ and 3″ diameter, to check an angle of 18° 30′, if the taper is measured from one side?

6. In grinding a reamer shank to fit a standard No. 2 Morse taper gage, it was found that the reamer lacked ⅜ of an inch of going into the gage to the gage mark. How much should be ground off of the diameter?

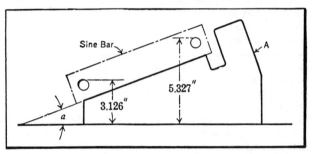

Fig. 2. Finding Angle *a* by means of a Sine Bar and Handbook Table

7. A milling machine arbor has a shank 6½ inches long with a No. 10 B. & S. taper. What is the total taper in this length?

8. A taper bushing for a grinding machine has a small inside diameter of ⅞″. It is 3″ long with ½″ taper per foot. Find the large inside diameter.

9. If a 5-inch sine bar is used for finding the angle of the tapering block *A* (Fig. 2) and the heights of the sine-bar plugs are as shown, find the corresponding angle *a* by means of the table beginning on Handbook page 1549.

SECTION 13

TOLERANCES AND ALLOWANCES FOR MACHINE PARTS

Handbook Pages 1506 to 1542

In manufacturing machine parts according to modern methods, certain maximum and minimum dimensions are established, particularly for the more important members of whatever machine or mechanism is to be constructed. These limiting dimensions serve two purposes: They prevent unnecessary accuracy and also excessive inaccuracies. A certain degree of accuracy is essential to the proper functioning of the assembled parts of a mechanism, but it is useless and wasteful to make parts more precise than needed to meet practical requirements; hence, the use of proper limiting dimensions promotes efficiency in manufacturing, and also insures standards of accuracy and quality that are consistent with the functions of the different parts of a mechanical device.

Parts made to specified limits usually are considered interchangeable or capable of use without selection but there are several degrees of interchangeability in machinery manufacture. Strictly speaking, interchangeability consists in making the different parts of a mechanism so uniform in size and contour that each part of a certain model will fit any mating part of the same model, regardless of the lot to which it belongs or when it was made. However, as often defined, interchangeability consists in making each part fit any mating part in a certain series; that is, the interchangeability exists only in the same series. Selective assembly is sometimes termed interchangeability, but it involves a selection or sorting of parts as explained later. It will be noted that the strict definition of interchangeability does not imply that the parts must always be assembled without hand work, although that is usually considered desirable. It does mean, however, that when the mating parts are finished, by whatever process, they must assemble and function properly, without fitting individual parts one to the other.

92

When a machine having interchangeable parts, has been installed possibly at some distant point, a broken part can readily be replaced by a new one sent by the manufacturer, but this feature is secondary as compared with the increased efficiency in manufacturing on an interchangeable basis. In order to make parts interchangeable, it is necessary to use gages and measuring tools, to provide some system of inspection, and to adopt suitable

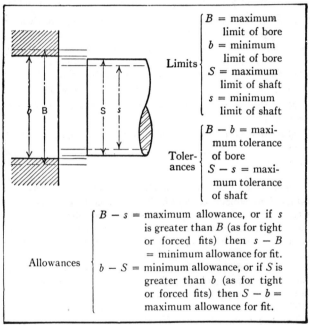

Limits
$\begin{cases} B = \text{maximum limit of bore} \\ b = \text{minimum limit of bore} \\ S = \text{maximum limit of shaft} \\ s = \text{minimum limit of shaft} \end{cases}$

Tolerances
$\begin{cases} B - b = \text{maximum tolerance of bore} \\ S - s = \text{maximum tolerance of shaft} \end{cases}$

Allowances
$\begin{cases} B - s = \text{maximum allowance, or if } s \\ \quad \text{is greater than } B \text{ (as for tight} \\ \quad \text{or forced fits) then } s - B \\ \quad = \text{minimum allowance for fit.} \\ b - S = \text{minimum allowance, or if } S \text{ is} \\ \quad \text{greater than } b \text{ (as for tight} \\ \quad \text{or forced fits) then } S - b = \\ \quad \text{maximum allowance for fit.} \end{cases}$

Fig. 1.   Diagram Showing Difference Between " Limit,"
" Tolerance," and "Allowance "

tolerances. Whether absolute interchangeability is practicable or not may depend upon the tolerances adopted, the relation between the different parts, and their form.

Meanings of the Terms " Limit," " Tolerance," and " Allowance." — The terms "limit" and "tolerance" and also "tolerance" and "allowance" are often used interchangeably, but these three terms each has a distinct meaning and refers to different dimensions. As shown by the accompanying diagram, Fig. 1, the *limits* of a hole or shaft are its diameters. *Tolerance* is the difference between two *limits* or limiting dimensions of a given part, and the term means that a certain amount of error is tolerated for practical reasons. *Allowance* is the difference between limiting dimensions on mating parts which are to be assembled either loosely or tightly, depending upon the amount allowed for the fit.

Example 1: — Limits and fits for cylindrical parts are given on pages 1515 to 1529 in the Handbook. This data provides a series of standard types and classes of fits. From the table on page 1521, establish limits of size and clearance for a 2-inch diameter hole and shaft for a class RC-1 fit (hole H5, shaft g4).

Max. hole = $2 + 0.0005 = 2.0005$; min. hole = $2 - 0 = 2$.
Max. shaft = $2 - .0004 = 1.9996$; min. shaft = $2 - 0.0007 = 1.9993$.
Min. allow. = min. hole − max. shaft = $2 - 1.9996 = 0.0004$.
Max. allow. = max. hole − min. shaft = $2.0005 - 1.9993 = 0.0012$.

Example 2: — Beginning on Handbook page 1303, there are tables of dimensions for the American Standard Thread Series — Class 3 Fit. Determine the pitch-diameter tolerance of both screw and nut, and also the minimum and maximum allowance between screw and nut at the pitch diameter, assuming that the nominal diameter is 1 inch and the pitch is 8 threads per inch.

The maximum pitch diameter or limit of the screw = 0.9188, and the minimum pitch diameter, 0.9134; hence, the tolerance = $0.9188 - 0.9134 = 0.0054$ inch.

The nut tolerance = $0.9242 - 0.9188 = 0.0054$ inch.

The maximum allowance for medium fit = maximum pitch diameter of nut − minimum pitch diameter of screw = $0.9242 - 0.9134 = 0.0108$ inch.

The minimum allowance = minimum pitch diameter of nut − maximum pitch diameter of screw = $0.9188 - 0.9188 = 0.0000$.

Relation of Tolerances to Limiting Dimensions and How Basic Size is Determined. — The absolute limits of the various dimensions and surfaces indicate danger points, inasmuch as parts made beyond these limits are unserviceable. A careful analysis of a mechanism shows that one of these danger points is more sharply defined than the other. For example, a certain stud must always assemble into a certain hole. If the stud is made beyond its maximum limit, it may be too large to assemble. If it is made beyond its minimum limit, it may be too loose or too weak to function. The absolute maximum limit in this case may cover a range of 0.001 inch, whereas the absolute minimum

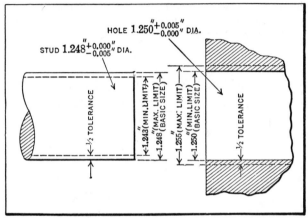

Fig. 2. Graphic illustration of the Meaning of the Term
Basic Size or Dimension

limit may have a range of at least 0.004 inch. In this case the maximum limit is the more sharply defined.

The basic size expressed on the component drawing is that limit which defines the more vital of the two danger points, while the tolerance defines the other. In general, the basic dimension of a male part such as a shaft, is the maximum limit which requires a minus tolerance. Similarly, the basic dimension of a female part is the minimum limit requiring a plus tolerance, as shown in Fig. 2. There are, however, dimensions which define neither a male nor a female surface, such for example as dimensions for the location of holes. In a few cases of this kind, a variation in one direction is less dangerous than a variation in the other. Under these conditions, the basic dimension represents the danger point, and the unilateral tolerance permits a variation only in the less dangerous direction. At other times, the conditions are such that any variation from a fixed point in either direction is equally dangerous. In such a case, the basic size represents this fixed point and tolerances on the drawing are bilateral and extend equally in both directions. (See Handbook page 1506 for explanation of unilateral and bilateral tolerances.)

When Allowance Provides Clearance Between Mating Parts. — When one part must fit freely into another part like a shaft in its bearing, the allowance between the shaft and bearing represents a clearance space. It is evident that the amount of clearance

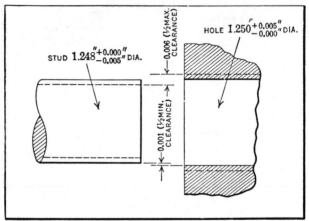

STUD $1.248^{+0.000''}_{-0.005''}$ DIA.

HOLE $1.250^{+0.005''}_{-0.000''}$ DIA.

0.006 (½ MAX. CLEARANCE)

0.001 (½ MIN. CLEARANCE)

Fig. 3.   Graphic Illustration of the Meaning of the Terms Maximum and Minimum Clearance

varies widely for different classes of work. The minimum clearance should be as small as will permit the ready assembly and operation of the parts, while the maximum clearance should be as great as the functioning of the mechanism will allow. The difference between the maximum and minimum clearances defines the extent of the tolerances. In general, the difference between the basic sizes of companion parts equals the minimum clearance (see Fig. 3), and the term "allowance," if not defined as maximum or minimum, is quite commonly applied to the minimum clearance.

When "Interference of Metal" is Result of Allowance. — If a shaft or pin is larger in diameter than the hole into which it is forced, there is, of course, interference between the two parts. The metal surrounding the hole is expanded and compressed as the shaft or other part is forced into place. Engine crankpins, car axles and various other parts are assembled in this way (see paragraph "Allowance for Forced Fits," Handbook page 1509). The force and shrink fits in Table 6 (pages 1527 and 1528) all represent interference of metal.

If interchangeable parts are to be forced together, the minimum interference establishes the danger point. This means

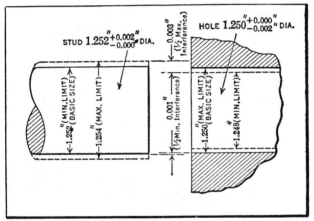

Fig. 4. Graphic Illustration of the Meaning of the Terms
Maximum and Minimum Interference

that for force fits the basic dimension of the shaft or pin is the minimum limit requiring a plus tolerance, while the basic dimension of the hole is the maximum limit requiring a minus tolerance. (See Fig. 4.)

Obtaining Allowance by Selection of Mating Parts. — The term "selective assembly" is applied to a method of manufacturing which is similar in many of its details to interchangeable manufacturing. In selective assembly, the mating parts are sorted according to size, and assembled or interchanged with little or no machining. The chief purpose of manufacturing by selective assembly, is the production of large quantities of duplicate parts as economically as possible. As a general rule, the smaller the tolerances, the more exacting and expensive will be the manufacturing processes; but it is possible to use comparatively large tolerances and then reduce them, in effect, by selective assembly, provided the quantity of parts is large enough to make such selective fitting possible. To illustrate the procedure, Fig. 5 shows a plug or stud which has a plus tolerance of 0.001 inch and a hole which also has a plus tolerance of 0.001 inch. Assume that this tolerance of 0.001 inch represents the normal size variation on each part when manufactured efficiently. With

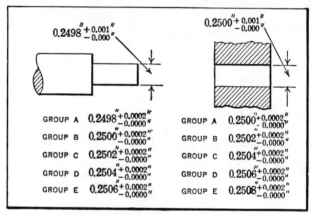

Fig. 5. Information Placed on Drawings Used in Selective Assembly Manufacturing to Facilitate Grading of Parts

this tolerance, a minimum plug in a maximum hole would have a clearance of $0.2510 - 0.2498 = 0.0012$ inch; and a maximum plug in a minimum hole would have a "metal interference" of $0.2508 - 0.2500 = 0.0008$ inch. But suppose the clearance required for these parts must range from zero to 0.0004 inch. This reduction can be obtained by dividing both plugs and holes into five groups. (See Fig. 5.) Any studs in Group A, for example, will assemble in any hole in Group A, but the studs in one group will not assemble properly in the holes in another group. When the largest stud in Group A is assembled in the smallest hole in Group A, the clearance equals zero. When the smallest stud in Group A is assembled in the largest hole in Group A, the clearance equals 0.0004 inch. Thus, in selective assembly manufacturing, there is a double set of limits, the first being the manufacturing limits, and the second the assembling limits. In many cases, two separate drawings are made of a part which is to be graded before assembly. One shows the manufacturing tolerances only, so as not to confuse the operator, while the other gives the proper grading information.

Example 3: — Data for force and shrink fits are given in the table on page 1527 in the Handbook. Establish the limits of size and interference of the hole and shaft for a Class FN-1 fit of 2-inch diameter.

Max. hole = 2 + 0.0007 = 2.0007; min. shaft = 2 − 0 = 2.

Max. shaft = 2 + 0.0018 = 2.0018; min. shaft = 2 + 0.0013 = 2.0013.

In the second column of the table the minimum and maximum interference are given as 0.0006 and 0.0018 inch, respectively, for a FN-1 fit of 2-inch diameter. For a "selected" fit, shafts are selected which are 0.0012 inch larger than the mating holes; that is, for any mating pair the shaft is larger than the hole by an amount midway between the minimum (0.0006 inch) and maximum (0.0018 inch) interference.

Dimensioning Drawings to Insure Obtaining Required Tolerances. — In dimensioning the drawings of parts requiring tolerances, there are certain fundamental rules that should be applied.

Rule 1. In interchangeable manufacturing there is only one dimension (or group of dimensions) in the same straight line which can be controlled within fixed tolerances. This is the distance between the cutting surface of the tool and the locating or registering surface of the part being machined. Therefore, it is incorrect to locate any point or surface with tolerances from more than one point in the same straight line.

Rule 2. Dimensions should be given between those points which it is essential to hold in a specific relation to each other. The majority of dimensions, however, are relatively unimportant in this respect. It is good practice to establish common location points in each plane and to give, as far as possible, all such dimensions from these points.

Rule 3. The basic dimensions given on component drawings for interchangeable parts should be, except for force fits and other unusual conditions, the "maximum metal" sizes (maximum shaft

or plug and minimum hole.)   The direct comparison of the basic
sizes should check the danger zone, which is the minimum clearance condition in the majority of cases.   It is evident that these
sizes are the most important ones, as they control the interchangeability, and they should be the first determined.   Once
established, they should remain fixed if the mechanism functions
properly and the design is unchanged.   The direction of the
tolerances, then, would be such as to recede from the danger zone.
In the majority of cases, this means that the direction of the
tolerances is such as will increase the clearance.   For force fits,

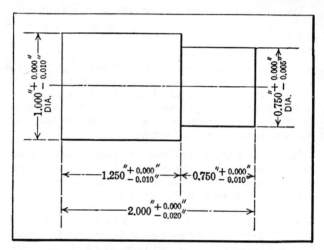

Fig. 6.   Common but Incorrect Method of Dimensioning

the basic dimensions determine the minimum interference, while
the tolerances limit the maximum interference.
Rule 4.   Dimensions must not be duplicated between the same
points.   The duplication of dimensions causes much needless
trouble, due to changes being made in one place and not in the
others.   It causes less trouble to search a drawing to find a
dimension than it does to have them duplicated and more readily
found but inconsistent.
Rule 5.   As far as possible, the dimensions on companion parts
should be given from the same relative locations.   Such a pro-

cedure assists in detecting interferences and other improper conditions.

In attempting to work in accordance with general laws or principles, one other elementary rule should always be kept in mind. Special cases require special consideration. The following detailed examples are given to illustrate the application of these five laws and to indicate results of their violation.

Violations of Rules for Dimensioning. — Fig. 6 shows a very common method of dimensioning a part such as the stud shown,

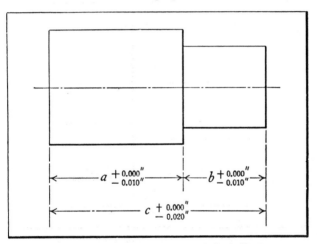

Fig. 7. One Interpretation of Dimensioning in Fig. 6

but one that is bad practice. It violates the first and second rules. As the dimensions given for the diameters are correct, they are eliminated from the discussion. The dimensions given for the various lengths are wrong: First, because they give no indication as to the essential lengths; second, because of several possible sequences of operations, some of which would not maintain the specified conditions.

Fig. 7 shows one possible sequence of operations indicated alphabetically. If we first finish the dimension a and then finish b, the dimension c will be within the specified limits. In this case, however, the dimension c is superfluous. Fig. 8 gives

another possible sequence of operations. If we first establish
a, and then b, the dimension c may vary 0.030 instead of 0.010
inch as is specified in Fig. 6. Fig. 9 gives a third possible sequence of operations. If we first finish the over-all length a,
and then the length of the body b, the stem c may vary 0.030
inch instead of 0.010 inch as specified in Fig. 6.

If three different plants were manufacturing this part, each
one using a different sequence of operations, it is evident from the
foregoing that a different product would be received from each

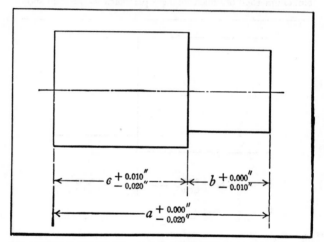

Fig. 8.   Another Interpretation of Dimensioning in Fig. 6

plant. The example given is the simplest one possible. As
the parts become more complex, and the number of dimensions
increase, the number of different combinations possible and the
extent of the variations in size that will develop also increase.

Fig. 10 shows the correct way to dimension this part if the
length of the body and the length of the stem are the essential
dimensions. Fig. 11 is the correct way if the length of the body
and the length over all are the most important. Fig. 12 is correct
if the length of the stem and the length over all are the most
important. If the part is dimensioned in accordance with
either Fig. 10, Fig. 11, or Fig. 12, the product from any number
of factories should be alike.

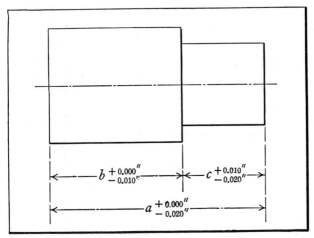

Fig. 9. A Third Interpretation of Dimensioning in Fig. 6

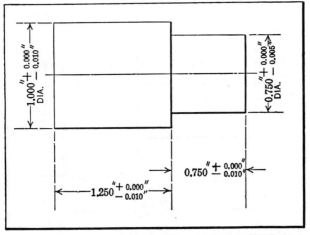

Fig. 10. Correct Dimensioning if Length of Body and Length
of Stem are Most Important

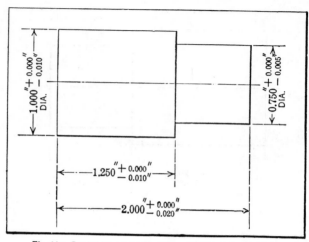

Fig 11. Correct Dimensioning if Length of Body and Over-all Length are Most Important

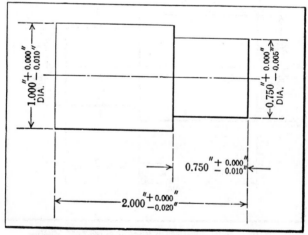

Fig. 12. Correct Dimensioning if Over-all Length and Length of Stem are Most Important

PRACTICE EXERCISES FOR SECTION 13

For answers to all practice exercise problems or questions
see Section 20

1. What factors influence the allowance for a forced fit?

2. What is the general practice in applying tolerances to center distances between holes?

3. A 2-inch shaft is to have a tolerance of 0.003 inch on the diameter. Show, by examples, three ways of expressing the shaft dimensions.

4. In what respect does a bilateral tolerance differ from a unilateral tolerance; show, by example?

5. What are the standard thickness tolerances for cold-rolled sheets?

6. What are the tolerances for hexagonal hot-rolled carbon steel bars?

7. Why do some manufacturers specify smaller tolerances for working gages than for inspection gages?

8. Name the different classes of fits for screw threads included in the American Standard.

9. How does the Unified Standard for Screw Threads differ from the former American Standard with regard to clearance between mating parts? With regard to working tolerance?

10. Under what conditions is one limiting dimension or "limit" also a basic dimension?

11. What do the letter symbols RC, LC, LT, LN and FN signify with regard to American Standard fits?

12. According to the table on the bottom of page 1490, broaching will produce work within tolerance grades 5 through 8. What does this mean in terms of thousandths of an inch, considering a 1-inch diameter broached hole?

SECTION 14

STANDARD SCREW AND PIPE THREADS

Handbook Pages 1255 to 1372

Different screw-thread forms and standards have been originated and adopted at various times, either because they were considered superior to other forms or because of the special requirements of screws used on a certain class of work.

A standard thread conforms to an adopted standard in regard to the form or contour of the thread itself, and as to the pitches or numbers of threads per inch for different screw diameters. A screw thread having either a modified form or a pitch which is either greater or less for a given screw diameter, than the adopted standard, is special.

The United States Standard formerly used in the United States was replaced by an American Standard having the same thread form as the former standard and a more extensive series of pitches, as well as tolerances and allowances for different classes of fits. This American Standard was revised in 1949 to include a Unified Thread Series which was established to obtain screw-thread interchangeability among the United Kingdom, Canada and the United States.

The Standard was revised again in 1959. The Unified threads are now the standard for use in the United States and the former American Standard threads are now used only in certain applications where the changeover in tools, gages, and manufacturing has not been completed. The differences between Unified and the former National Standard threads are explained on pages 1255 and 1257 in the Handbook.

As may be seen in the table on Handbook page 1261, the Unified Series of screw threads consists of three standard series having graded pitches (UNC, UNF, and UNEF) and eight standard series of uniform (constant) pitch. In addition to these standard series, there are places in the table beginning on page 1276 where special threads (UNS) are listed. These UNS threads are for use only if standard series threads do not meet requirements.

Example 1: — The table on Handbook page 1262 shows that the pitch diameter of a 2-inch screw thread is 1.8557 inches. What is meant by the term "pitch diameter" as applied to a screw thread and how is it determined?

According to a definition of "pitch diameter" given in connection with American Standard screw threads, the pitch diam-

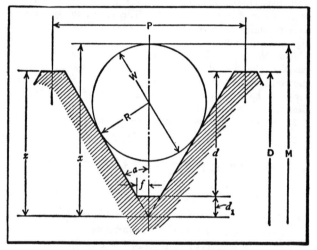

Fig. 1. Diagram Illustrating the Derivation of Formulas for Three-wire
Measurements of Screw Thread Pitch Diameters

eter of a straight (non-tapering) screw thread is the diameter of an imaginary cylinder the surface of which would pass through the threads at such points as to make equal the width of the threads and the width of the spaces cut by the surface of the cylinder.

The basic pitch diameter equals the basic major (outside) diameter minus two times the addendum of the external thread (page 1108). In this case, the basic pitch diameter is 2.0000 − 2 × 0.07217 = 1.8557 inches.

Example 2: — According to the table on Handbook page 1159, the tensile strength of a United States Standard bolt $3\frac{1}{2}$ inches in diameter, at a stress of 6000 pounds per square inch, is 45,300 pounds. How is this figure determined?

The diameter at the root of the thread or where the bolt is weakest, is 3.1 inches; hence the area is 7.5477 square inches If the safe stress is limited to 6000 pounds per square inch, then the allowable load for an area of 7.5477 square inches = 7.5477 × 6000 = 45,300 pounds approximately.

Note that the working strength given by the second **formula** (see page 1157) is 40,036 pounds for a $3\frac{1}{2}$-inch bolt and a stress of 6000 pounds per square inch. This reduction, as compared with the table on page 1159, is to allow for the stress due to tightening the nut.

Example 3:—On Handbook page 1379 formulas are given for checking the pitch diameter of screw threads by the three-wire method (when effect of lead angle is ignored). Show how these formulas have been derived, using the one for the American Standard thread as an example.

It is evident from the diagram, Fig. 1, that

$$M = D - 2z + 2x \tag{1}$$

$$x = R + \frac{R}{\sin a} \quad \text{and} \quad 2x = 2R + \frac{2R}{0.5}; \text{ hence}$$

$$2x = \frac{(2 \times 0.5 + 2)R}{0.5} = \frac{3R}{0.5} = 6R = 3W$$

$$z = d + d_1 = 0.6495P + f \times \cot a$$

$$f = 0.0625P; \text{ therefore}$$

$$z = 0.6495P + 0.10825P = 0.75775P$$

If in formula (1) we substitute the value of $2z$ or $2 \times 0.75775P$ and the value of $2x$, we have

$$M = D - 1.5155 \times P + 3W \tag{2}$$

This formula (2) is the one found in the Handbook. If the measurement M over the wires does not equal $D - 1.5155 \times P + 3W$ this shows that the pitch diameter of the screw thread is incorrect.

Example 4: — On Handbook page 1387 is given a formula for checking the angle of a screw thread by a three-wire method. How is this formula derived? Referring to the diagram, Fig. 2

$$\sin a = \frac{W}{S} \tag{1}$$

If D = diameter of larger wires and d = diameter of smaller wires

$$W = \frac{D - d}{2}$$

If B = difference in measurement over wires then

$$S = \frac{B - (D - d)}{2}$$

By inserting these expressions for W and S in formula (1) and

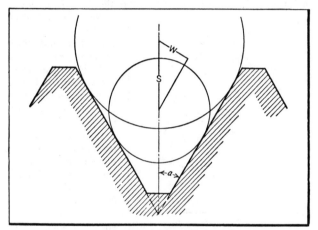

Fig. 2. Diagram Illustrating the Derivation of Formula for Checking the Thread Angle by the Three-wire System

cancelling, the formula given in the Handbook is obtained if A is substituted for $D - d$.

$$\sin a = \frac{A}{B - A} \qquad (2)$$

Example 5: — A vernier gear-tooth caliper (like the one shown on Handbook page 760) is to be used for checking the width of an Acme screw by measuring squarely across or perpendicular to the thread. Since standard screw thread dimensions are in the plane of the axis, how is the width square or normal to the sides of the thread determined? Assume that the width is to be measured at the pitch line and that the number of threads per inch is two.

The table on Handbook page 1320 shows that for two threads per inch, the depth is 0.260 inch; hence, if the measurement is to be at the pitch line, the vertical scale of the caliper is set to (0.260 − 0.010) ÷ 2 = 0.125 inch. The pitch equals

$$\frac{1}{\text{No. of Threads per Inch}} = \tfrac{1}{2} \text{ inch}$$

The width A, Fig. 3, in the plane of the axis equals $\frac{1}{2}$ of the pitch, or $\frac{1}{4}$ inch. The width B perpendicular to the sides of the thread

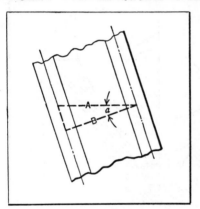

= width in axial plane × cosine helix angle. (The helix angle which equals angle a, is based upon the pitch diameter and is measured from a plane perpendicular to the axis of the screw thread.) The width A in the plane of the axis represents the hypotenuse of a right triangle, and the required width B equals the side adjacent; hence width B = A × cosine of helix angle. The angle of the thread itself (29° for an

Fig. 3.   Determining the Width Perpendicular to the Sides of a Thread at the Pitch Line

Acme Thread) does not affect the solution.

Width of Flat End of Unified Screw Thread and American Standard Acme Screw Thread Tools. — The widths of flat or end of the threading tool for either of these threads may be measured by using a micrometer as illustrated at A, Fig. 4. In measuring the thread tool, a scale is held against the spindle and anvil of the micrometer and the end of the tool is placed against this scale. The micrometer is then adjusted to the position shown and 0.2887 inch subtracted from the reading for an American Standard screw thread tool and for the American Standard Acme threads, 0.1293 inch is subtracted from the micrometer reading to obtain the width of the tool point. The constants (0.2887 and 0.1293) which are subtracted from the micrometer reading are only correct when the micrometer spindle has the usual diameter of 0.25 inch.

An ordinary gear-tooth vernier caliper may also be used for testing the width of a thread tool point, as illustrated at B. If the measurement is made at a vertical distance x of $\frac{1}{4}$ inch from the points of the caliper jaws, the constants previously given for American Standard and American Standard Acme threads should be subtracted from the caliper reading to obtain the actual width of the cutting end of the tool.

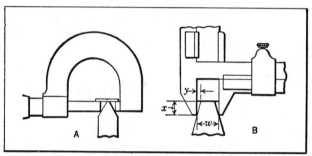

Fig. 4.   Measuring Width of Flat on Threading Tool (A) with a
Micrometer; (B) with a Gear Tooth Vernier

Example 6: — Explain how the constants 0.2887 and 0.1293 referred to in a preceding paragraph, are derived and deduce a general rule applicable regardless of the micrometer spindle diameter or vertical dimension x, Fig. 4.

The dimension x (which also is equivalent to the micrometer spindle diameter) represents one side of a right triangle (the side adjacent), having an angle of $29 \div 2 = 14$ degrees and 30 minutes, in the case of an Acme thread. The side opposite, or $y =$ side adjacent $\times$ tangent = dimension $x \times$ tan 14° 30'.

If x equals 0.25 inch, then side opposite or $y = 0.25 \times 0.25862 = 0.06465$; hence, the caliper reading minus 2 $\times$ 0.06465 = width of the flat end (2 $\times$ 0.06465 = 0.1293 = constant).

The same result would be obtained by multiplying 0.25862 by $2x$; hence, the following rule: To determine the width of the end of the threading tool, by the general method illustrated in Fig. 4, multiply twice the dimension x (or spindle diameter in the case of the micrometer) by the tangent of one-half the thread tool angle, and subtract this product from the width w to obtain the width at the end of the tool.

Example: — A gear tooth vernier caliper is to be used for measuring the width of the flat of an American Standard external screw thread tool. The vertical scale is set to $\frac{1}{8}$ inch (corresponding to the dimension x, Fig. 4). How much is subtracted from the reading on the horizontal scale, to obtain the width of the flat end of the tool?

$$\tfrac{1}{8} \times 2 \times \tan 30° = \tfrac{1}{4} \times 0.57735 = 0.1443 \text{ inch}$$

Hence, the width of the flat equals w, Fig. 4 minus 0.1443. This width should be equal to one-eighth of the pitch of the thread to be cut, since this is the width of flat at the minimum minor diameter of American Standard external screw threads.

PRACTICE EXERCISES FOR SECTION 14

For answers to all practice exercise problems or questions
see Section 20

1. What form of screw thread is most commonly used (*a*) in the United States?   (*b*) in England?

2. What is the meaning of the abbreviation 3″ — 4NC–2?

3. What are the advantages of an Acme thread as compared with a square thread?

4. For what reason would a Stub Acme thread be preferred in some applications?

5. Find the pitch diameters of the following screw threads of American Standard form: $\frac{1}{4}$ — 28, (meaning $\frac{1}{4}$ inch diameter and 28 threads per inch); $\frac{3}{4}$ — 10?

6. How much taper has a standard pipe thread?

7. Under what conditions are straight, or non-tapering pipe threads used?

8. In cutting a taper thread, what is the proper position for the lathe tool?

9. If a lathe is used for cutting a British Standard pipe thread, in what position is the tool set?

10. A thread tool is to be ground for cutting an Acme thread having 4 threads per inch; what is the correct width of the tool at the end?

11. What are the common shop and tool-room methods of checking the pitch diameters of American Standard screw threads requiring accuracy?

12. In using the formula, page 1379, for measuring an American Standard screw thread by the three-wire method, why should the constant 1.5155 be multiplied by the pitch before subtracting from diameter D, even if not enclosed by parentheses?

13. What is the difference between the pitch and the lead (a) of a double thread?   (b) of a triple thread?

14. In using a lathe to cut American Standard threads, what should be the truncations of the tool points and the thread depths for the following pitches: 0.1, 0.125, 0.2 and 0.25 inch?

15. In using the three-wire method of measuring a screw thread, what is the micrometer reading for a $\frac{3}{4} - 12$ special thread of American Standard form, if the wires have a diameter of 0.070 inch?

16. Are most nuts made to the United States Standard dimensions as given on Handbook page 1159?

17. Is there, at the present time, a Manufacturers Standard for bolts and nuts?

18. The American Standard for machine screws includes a coarse-thread series and a fine-thread series as shown by the tables on Handbook pages 1262 and 1263. Which series is commonly used?

19. How is the length (a) of a flat-head or countersunk type of machine screw measured?   (b) of a fillister head machine screw?

20. What size tap drill should be used for an American Standard machine screw of No. 10 size, 24 threads per inch?

21. What is the diameter of a No. 10 drill?

22. Is a No. 6 drill larger than a No. 16?

23. What is the relation between the letter size drills and the numbered sizes?

24. Why is it common practice to use tap drills that leave about ¾ of the full thread depth after tapping, as shown by the tables on pages 1405 and 1406?

25. What form of screw thread is used on (a) machine screws? (b) cap screws?

26. What standard governs the pitches of cap-screw threads?

27. What form of thread is used on the National Standard fire hose couplings? How many standard diameters are there?

28. In what way do hand taps differ from machine screw taps?

29. What are tapper taps?

30. The diameter of a ¾ — 10 American Standard thread is to be checked by the three-wire method. What is the largest size wire that can be used?

31. Why is the advance of some threading dies positively controlled by a lead-screw instead of relying upon the die to lead itself?

32. What type of die is adapted to comparatively long taper threads?

33. What is the included angle of the heads of American Standard (a) flat-head machine screws? (b) flat-head cap-screws? (c) flat-head wood screws?

SECTION 15

PROBLEMS IN MECHANICS

Handbook Pages 289 to 356

In the design of machines or other mechanical devices, it is often necessary to deal with the action of forces and their effect. For example, the problem may be to determine what force is equivalent to two or more forces acting in the same plane but in different directions. Another type of problem is to determine the change in the magnitude of a force resulting from the application of mechanical appliances such as levers, pulleys and screws used either separately or in combination. It may also be necessary to determine the magnitude of a force in order to proportion machine parts to resist the force safely; or, possibly, to ascertain if the force is great enough to perform a given amount of work. Determining the energy stored in a moving body or its capacity to perform work, and the power developed by mechanical apparatus, or the rate at which work is performed, are additional examples of problems frequently encountered in originating or developing mechanical appliances. That section in MACHINERY'S HANDBOOK on Mechanics, beginning page 289, deals with fundamental principles and formulas applicable to a large variety of problems of the general classes referred to.

The Moment of a Force. — The tendency of a force acting upon a body is, in general, to produce either a motion of translation (that is, to cause every part of the body to move in a straight line) or to produce a motion of rotation. A *moment*, in mechanics, is the measure of the turning effect of a force which tends to produce rotation. For example, suppose a force acts upon a body which is supported by a pivot. Unless the line of action of the force happens to pass through the pivot, the body will tend to rotate. Its tendency to rotate, moreover, will depend upon two things: (1) upon the magnitude of the force acting, and (2) upon the distance of the force from the pivot, *measuring along a line at right angles to the line of action of the force.* (See the diagram at the bottom of Handbook page 296, and the accompanying text.)

115

Example 1: — A force F of 300 pounds is applied to a crank disk A (Fig. 1) and in the direction of the arrow. If the radius $R = 5$ inches, what is the turning moment? Also determine how much the turning moment is reduced when the crankpin is in the position shown by the dotted lines, assuming that the force is along line f and that $r = 2\frac{1}{2}$ inches.

When the crankpin is in the position shown by the full lines, the maximum turning moment is obtained and it equals $F \times R$ $= 300 \times 5 = 1500$ inch-pounds or pound-inches. When the crankpin is in the position shown by the dotted lines, the turning moment is reduced one-half and equals $f \times r = 300 \times 2\frac{1}{2}$ $= 750$ inch-pounds.

Note: Torque or turning moment is sometimes expressed as pound-feet or pound-inches, instead of using the term foot-

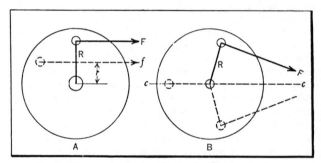

Fig. 1. Diagram showing how the Turning Moment of a Crank Disk Varies from Zero to Maximum

pounds or inch-pounds. Since *foot-pound* is the unit of work and is used in horsepower calculations, it is considered preferable to reverse it and use the term *pound-foot* to indicate torque or turning moment. The reversal of the term foot-pound serves to distinguish readily between the two units of measurement — the unit of work and the unit of turning moment. The latter ordinarily is expressed as inch-pounds or pound-inches instead of foot-pounds or pound-feet, because the dimensions of shafts and other machine parts ordinarily are given in inches; hence the reversal of the term *inch-pound* is not so important as in the case of foot-pound, because inch-pound is not used as a unit of work.

Example 2: — Assume that the force F (diagram B, Fig. 1) is applied to the crank through a rod connecting with a crosshead which slides along center-line c–c. If the crank radius $R = 5$ inches, what will be the maximum and minimum turning moments?

The maximum turning moment occurs when the radial line R is perpendicular to the force line F and equals in inch-pounds, $F \times 5$ in this example. When the radial line R is in line with the center-line c–c, the turning moment is 0, since $F \times 0 = 0$.

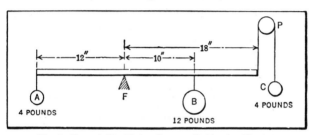

Fig. 2. Lever in Equilibrium Because the Turning Moment
of Opposing Forces are Equal

This is the "dead-center" position for steam engines and explains why the crankpins on each side of a locomotive are located 90 degrees apart, or, in such a position that the maximum turning moment, approximately, occurs when the turning moment is zero on the opposite side. With this arrangement, it is always possible to start the locomotive since only one side at a time can be in the dead-center position.

The Principle of Moments in Mechanics. — When two or more forces act upon a rigid body and tend to turn it about an axis, then, for equilibrium to exist, the sum of the moments of the forces which tend to turn the body in one direction must be equal to the sum of the moments of those which tend to turn it in the opposite direction about the same axis.

Example 3: — In Fig. 2, a lever 30 inches long is pivoted at the fulcrum F. At the right, and 10 inches from F is a weight, B, of 12 pounds tending to turn the bar in a right-hand direction about its fulcrum F. At the left end, 12 inches from F, the weight A of 4 pounds tends to turn the bar in a left-hand direction, while weight C, at the other end, 18 inches from F, has a like effect, through the use of the string and pulley P. Taking moments about F, which is the center of rotation, we have:

Moment of B = 10 × 12 = 120 inch-pounds

Opposed to this are the moments of A and C:

Moment of A = 4 × 12 = 48 inch-pounds
Moment of C = 4 × 18 = 72 inch-pounds

Sum of negative moments = 120 inch-pounds

Hence, the opposing moments are equal, and, if we suppose, for simplicity, that the lever is weightless, it will balance or be in equilibrium. Should weight A be increased, the negative

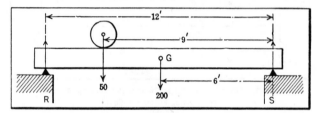

Fig. 3. The Weight on Each Support is Required

moments would be greater, and the lever would turn to the left, while if B should be increased or its distance from F be made greater, the lever would turn to the right. (See fourth diagram on Handbook page 296 and accompanying text.)

Example 4: — Another application of the principle of moments is given in Fig. 3. A beam of uniform cross section, weighing 200 pounds, rests upon two supports, R and S, which are 12 feet apart. The weight of the beam is considered to be concentrated at its center of gravity G, at a distance of 6 feet from each support. A weight of 50 pounds is placed upon the beam at a distance of 9 feet from the right-hand support S. Required, the portion of the total weight borne by each support.

Before proceeding, it should be explained that the two supports react or push upward, with a force equal to the downward pressure of the beam. To make this clear, suppose two men to take hold of the beam, one at each end, and that the supports be withdrawn. Then, in order to hold the beam in position, the two men must together lift or pull upward an amount equal to the weight of the beam and its load, or 250 pounds. Placing the supports in position again, and resting the beam upon them, does not change the conditions. The weight of the beam acts downward, and the supports react by an equal amount.

Now, to solve the problem, assume the beam to be pivoted at one support, say at S. The forces or weights of 50 pounds and 200 pounds tend to rotate the beam in a left-hand direction about this point, while the reaction of R in an upward direction tends to give it a right-hand rotation. As the beam is balanced and has no tendency to rotate, it is in equilibrium, and the opposing moments of these forces must balance; hence, taking moments,

$$9 \times 50 = 450 \text{ pound-feet}$$
$$6 \times 200 = 1200 \text{ pound-feet}$$

Sum of negative moments = 1650 pound-feet

Letting R represent the reaction of support,

Moment of $R = R \times 12$ = pound-feet

By the principle of moments, $R \times 12 = 1650$. That is, if R, the quantity which we wish to obtain, be multiplied by 12, the result will be 1650; hence, to obtain R, divide 1650 by 12, whence $R = 137.5$ pounds, which is also the weight of that end of the beam. As the total load is 250 pounds, the weight at the other end must be $250 - 137.5 = 112.5$ pounds.

The Principle of Work in Mechanics — There is another principle of more importance than the principle of moments, even in the study of machine elements. It is called the principle of work. According to this principle (neglecting frictional or other losses) the applied force, multiplied by the distance through which it moves, equals the resistance overcome, multiplied by the distance through which it is overcome. The principle of work may also be stated as follows:

Work put in = lost work + work done by machine

This principle holds absolutely in every case. It applies equally to a simple lever, the most complex mechanism, or to a so-called "perpetual motion" machine. No machine can be made to perform work unless a somewhat greater amount — enough to make up for the losses — be applied by some external agent. In the "perpetual motion" machine no such outside force is supposed to be applied, hence such a machine is impossible, and against all the laws of mechanics.

Example 5: — Assume that a rope exerts a pull F of 500 pounds (upper diagram, Handbook page 308) and that the pulley radius $R = 10$ inches and the drum radius $r = 5$ inches. How much weight W can be lifted (ignoring frictional losses) and upon what mechanical principle is the solution based?

According to one of the formulas accompanying the diagram at the top of Handbook page 308,

$$W = \frac{F \times R}{r} = \frac{500 \times 10}{5} = 1000 \text{ pounds}$$

This formula (and the others for finding the values of F, R, etc.) agrees with the principle of moments, and also with the principle of work. The principle of moments will be applied first.

The moment of the force F about the center of the pulley, which corresponds to the fulcrum of a lever, is F multiplied by the perpendicular distance R, it being a principle of geometry that a radius is perpendicular to a line drawn tangent to a circle, at the point of tangency. Also the opposing moment of W is $W \times r$. Hence, by the principle of moments,

$$F \times R = W \times r$$

Now, for comparison, we will apply the principle of work. Assuming this principle to be true, force F multiplied by the distance traversed by this force or by a given point on the rim of the large pulley, should equal the resistance W multiplied by the distance that the load is raised. In one revolution force F passes through a distance equal to the circumference of the pulley which is equal to $2 \times 3.1416 \times R = 6.2832 \times R$, and the hoisting rope passes through a distance equal to $2 \times 3.1416 \times r$. Hence, by the principle of work,

$$6.2832 \times F \times R = 6.2832 \times W \times r$$

This statement simply shows that $F \times R$ multiplied by 6.2832 equals $W \times r$ multiplied by the same number, and it is evident therefore, that the equality will not be altered by canceling the 6.2832 and writing

$$F \times R = W \times r$$

But this is the same statement obtained by applying the principle of moments; hence, we see that the principle of moments and the principle of work harmonize.

The basis of operation of a train of wheels is a continuation of the principle of work. For example, in the gear train represented by the diagram at the bottom of Handbook page 308, the continued product of the applied force F and the radii of the driven wheels equals the continued product of the resistance W and the radii of the drivers. In calculations, the pitch diameters, or the number of teeth in gear wheels may be used instead of the radii.

Efficiency of a Machine or Mechanism. — The efficiency of a machine is the ratio of the power delivered by the machine to the power received by it. For example, the efficiency of an electric motor is the ratio between the power delivered by the motor to the machinery which it drives, and the power it receives from the generator. Assume, for example, that a motor receives 50 kilowatts from the generator, but that the output of the motor is only 47 kilowatts. Then, the efficiency of the motor is $47 \div 50 = 94$ per cent. The efficiency of a machine tool is the ratio of the power consumed at the cutting tool to the power delivered by the driving belt. The efficiency of gearing is the ratio between the power obtained from the driven shaft to the power used by the driving shaft. Generally speaking, the efficiency of any machine or mechanism is the ratio of the "output" of power to the "input." The percentage of power representing the difference between the "input" and "output" has been, dissipated through frictional and other mechanical losses.

Mechanical Efficiency. — If E represents the energy which a machine transforms into useful work or delivers at the driven end; L equals the energy lost through friction or dissipated in other ways; then

$$\text{Mechanical Efficiency} = \frac{E}{E + L}$$

In this case the total energy $E + L$ is assumed to be the amount that is transformed into useful and useless work. The actual total amount of energy, however, may be considerably larger than the amount represented by $E + L$. For example, in a steam engine there are heat losses due to radiation and steam condensation, and considerable heat energy supplied to an internal combustion engine is dissipated either through the cooling water or direct to the atmosphere. In other classes of mechanical and electrical machinery the total energy is much larger than that represented by the amount transformed into useful and useless work.

Absolute Efficiency. — If E_1 equals the full amount of energy or the true total, then

$$\text{Absolute Efficiency} = \frac{E}{E_1}$$

It is evident that absolute efficiency of a prime mover, such as a steam or gas engine, will be much lower than the mechanical efficiency. Ordinarily, the term efficiency as applied to engines and other classes of machinery, means the mechanical efficiency.

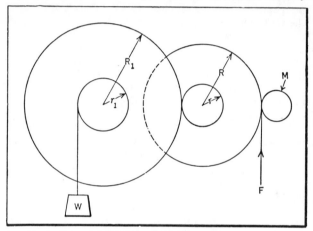

Fig. 4. Determining the Power Required for Lifting a Weight by Means of Motor and Compound Train of Gearing

The mechanical efficiency of reciprocating steam engines may vary from 85 to 95 per cent, but the *thermal* efficiency may range from 5 to 25 per cent, the smaller figure representing non-condensing engines of the cheaper class and the higher figure the best types.

Example 6: — Assume that a motor driving through a compound train of gearing (see diagram, Fig. 4) is to lift a weight W of 1000 pounds. The pitch radius $R = 6$ inches; $R_1 = 8$ inches; pitch radius of pinion $r = 2$ inches; and radius of winding drum $r_1 = 2\frac{1}{2}$ inches. What motor horsepower will be required if the frictional loss in the gear train and bearings is assumed to be

10 per cent? The pitch-line velocity of the motor pinion M is 1200 feet per minute.

The problem is to determine first the tangential force F required at the pitch line of the motor pinion; then the equivalent horsepower is easily found. According to the formula at the bottom of the Handbook, page 308, which does not take into account frictional losses

$$F = \frac{1000 \times 2 \times 2\frac{1}{2}}{6 \times 8} = 104 \text{ pounds}$$

The pitch-line velocity of the motor pinion is 1200 feet per minute and as the friction loss is assumed to be 10 per cent, the mechanical efficiency equals $90 \div (90 + 10) = 0.90$ or 90 per cent as commonly written; hence

$$\text{Horsepower} = \frac{104 \times 1200}{33,000 \times 0.90} = 4\frac{1}{4} \text{ approximately}$$

Example 7: — In actual designing practice, a 5 horsepower motor, or one of larger size, might be selected for the drive referred to in Example 6 (depending upon conditions), to provide extra power in case it is needed. Assume, however, in order to illustrate the procedure, that the gear train is to be modified so that the calculated horsepower will be 4 instead of $4\frac{1}{4}$; conditions otherwise are the same as in Example 6.

$$F = \frac{33,000 \times 4}{1200} = 110 \text{ pounds}$$

Hence since $W = 1000$ pounds,

$$1000 = \frac{110 \times 0.90 \times R \times R_1}{r \times r_1}$$

Insert any values for the pitch radii R, R_1, etc., that will balance the equation, so that the right-hand side equals 1000, at least approximately. Several trial solutions may be necessary to obtain a total of about 1000 and at the same time, secure properly proportioned gears that meet other requirements of the design. In this case, suppose the same radii are used, except R_1, which is increased from 8 to $8\frac{1}{2}$ inches. Then

$$\frac{110 \times 0.90 \times 6 \times 8\frac{1}{2}}{2 \times 2\frac{1}{2}} = 1000 \text{ approximately}$$

This shows that the increase in the radius of the last driven gear from 8 to $8\frac{1}{2}$ inches makes it possible to use the four horsepower motor. The hoisting speed has been decreased somewhat and the center distance between the gears has been increased. **These** changes might or might not be objectionable in actual designing practice, depending upon the particular requirements.

Force Required to Turn a Screw Used for Elevating or Lowering Loads. — In determining the force which must be applied at the end of a given lever-arm in order to turn a screw (or nut surrounding it), there are two conditions to be considered: (1) When rotation is such that the load *resists* the movement of the screw, as in raising a load with a screw jack; (2) when rotation is such that the load *assists* the movement of the screw, as in lowering a load. The formulas at the bottom of the table on Handbook page 309, apply to both of these conditions. When the load resists the screw movement, use the formula "for motion in a direction opposite to Q." When the load assists the screw movement, use the formula "for motion in the same direction as Q."

If the lead of the thread is large in proportion to the diameter so that the helix angle is large, the force F may have a negative value, which indicates that the screw will turn due to the load alone, unless resisted by a force which is great enough to prevent rotation of a non-locking screw.

Example 8: — A screw is to be used for elevating a load Q of 6000 pounds. The pitch diameter is 4 inches, the lead is 0.75 inch and the coefficient of friction between screw and nut is assumed to be 0.150. What force F will be required at the end of a lever arm R of 10 inches? In this example, the load is in the direction opposite to arrow Q (see diagram at bottom of the table on Handbook page 309).

$$F = 6000 \times \frac{0.75 + 6.2832 \times 0.150 \times 2}{6.2832 \times 2 - 0.150 \times 0.75} \times \frac{2}{10} = 254 \text{ pounds}$$

Example 9: — What force F will be required to lower a load of 6000 pounds using the screw referred to in Example 8? In this case, the load assists in turning the screw; hence

$$F = 6000 \times \frac{6.2832 \times 0.150 \times 2 - 0.75}{6.2832 \times 2 + 0.150 \times 0.75} \times \frac{2}{10} = 108 \text{ pounds}$$

Coefficients of Friction for Screws and their Efficiency. — According to experiments by Professor Kingsbury made with square-threaded screws, a coefficient of 0.10 is about right for pressures less than 3000 pounds per square inch and velocities above 50 feet per minute, assuming that fair lubrication is maintained. If the pressures vary from 3000 to 10,000 pounds per square inch, a coefficient of 0.15 is recommended for low velocities. The coefficient of friction varies according to lubrication and the materials used for the screw and nut. For pressures of 3000 pounds per square inch and using heavy machinery oil as a lubricant, the coefficients were as follows: Mild steel screw and cast-iron nut, 0.132; mild steel nut, 0.147; cast brass nut, 0.127. For pressures of 10,000 pounds per square inch using a mild steel screw, the coefficients were, for a cast-iron nut, 0.136; for a mild steel nut, 0.141; for a cast brass nut, 0.136. For dry screws, the coefficient may be 0.3 to 0.4 or higher.

Frictional resistance is proportional to the normal pressure, and for a thread of angular form, the increase in the coefficient of friction is equivalent practically to $\mu \sec \beta$, in which β equals one-half the included thread angle; hence, for a sixty-degree thread, a coefficient of 1.155μ may be used. The square form of thread has a somewhat higher efficiency than threads with sloping sides, although when the angle of the thread form is comparatively small, as in the case of an Acme thread, there is little increase in frictional losses. Multiple-thread screws are much more efficient than single-thread screws, as the efficiency is affected by the helix angle of the thread.

The efficiency between a screw and nut increases quite rapidly for helix angles up to 10 or 15 degrees (measured from a plane perpendicular to the screw axis). The efficiency remains nearly constant for angles between about 25 and 65 degrees, and the angle of maximum efficiency is between 40 and 50 degrees. A screw will not be self-locking if the efficiency exceeds 50 per cent. For example, the screw of a jack or other lifting or hoisting appliance would turn under the action of the load if the efficiency were over 50 per cent. It is evident that maximum efficiency for power transmission screws often is impracticable, as for example, when the smaller helix angles are required to permit moving a given load by the application of a smaller force or turning moment than would be needed for a mutiple screw thread.

In determining the efficiency of a screw and a nut, the helix angle of the thread and the coefficient of friction are the important factors. If E equals the efficiency, A equals the helix angle, measured from a plane perpendicular to the screw axis, and μ equals the coefficient of friction between the screw thread and nut, then the efficiency may be determined by the following formula, which does not take into account any additional friction losses, such as may occur between a thrust collar and its bearing surfaces:

$$E = \frac{\tan A \,(1 - \mu \tan A)}{\tan A + \mu}$$

This formula would be suitable for a screw having ball-bearing thrust collars. Where collar friction should be taken into account, a fair approximation may be obtained by changing the denominator of the foregoing formula to $\tan A + 2\,\mu$. Otherwise the formula remains the same.

Angles and Angular Velocity Expressed in Radians. — There are three systems generally used to indicate the size of angles. They are ordinarily measured by the number of degrees in the arc subtended by the sides of the angle. Thus if the arc subtended by the sides of the angle equals one-sixth of the circumference, the angle is said to be 60 degrees. Angles are also designated as multiples of a right angle. As an example of this, the sum of the interior angles of any polygon equals the number of sides less two, times two right angles. Thus the sum of the interior angles of an octagon equals $(8 - 2) \times 2 \times 90 = 6 \times 180 = 1080$ degrees. Hence each interior angle equals $1080 \div 8 = 135$ degrees.

A third method of designating the size of an angle is very helpful in certain problems. This method makes use of radians. A radian is defined as a central angle, the subtended arc of which equals the radius of the arc.

Using the symbols on Handbook page 353, v may represent the length of an arc as well as the velocity of a point on the periphery of a body. Then according to the definition of a radian: $\omega = \dfrac{v}{r}$ or, the angle in radians equals the length of the arc divided by the radius. Both the length of the arc and the radius must, of course, have the same unit of measurement — both must be in feet or inches or centimeters, etc. By rearranging the preceding equation:

$$v = \omega r \quad \text{and} \quad r = \frac{v}{\omega}$$

These three formulas will solve practically every problem involving radians.

The circumference of a circle equals πd or $2\pi r$ which equals 6.2832 r. This indicates that a radius is contained in a circumference 6.2832 times; hence there are 6.2832 radians in a circumference. Since a circumference represents 360 degrees, one radian equals 360 ÷ 6.2832 = 57.2958 degrees. Since 57.2958 degrees = 1 radian, 1 degree = 1 radian ÷ 57.2958 = 0.01745 radian.

Example 10: — 2.5 radians equal how many degrees?   One radian = 57.2958 degrees; hence, 2.5 radians = 57.2958 × 2.5 = 143.239 degrees.

Example 11: — 22° 31′ 12″ = how many radians?   12 seconds $= \frac{12}{60} = \frac{1}{5} = 0.2$ minute; $31.2′ \div 60 = 0.52$ degree.   One radian = 57.3 degrees approximately. $22.52° = 22.52 \div 57.3 = 0.393$ radian.

Example 12: — In the figure on Handbook page 72, Let $l = v = 30$ inches; $r = 50$ inches; find the central angle. $\omega = \dfrac{v}{r} = \dfrac{30}{50} = \dfrac{3}{5} = 0.6$ radian.

$$57.2958 \times 0.6 = 34° 22.6′$$

Example 13: — $\dfrac{3\pi}{4}$ radians equal how many degrees? 2π radians = 360°; π radians = 180°. $\dfrac{3\pi}{4} = \dfrac{3}{4} \times 180 = 135$ degrees

Example 14: — A 20-inch grinding wheel has a surface speed of 6000 feet per minute.   What is the angular velocity?

The radius $(r) = \frac{10}{12}$ feet; the velocity (v) in feet per second $= \frac{6000}{60}$; hence,

$$\omega = \frac{6000}{60 \times \frac{10}{12}} = 120 \text{ radians per second}$$

Example 15: — Use table on page 275 to solve Example 11.

22°	= 0.349066 radians	
2°	= 0.034907	"
31′	= 0.009018	"
12″	= 0.000058	"
22° 31′ 12″	= 0.393049	"

Example 16: — 7.23 radians equals how many degrees?
On Handbook page 276 find:

7.0	radians	=	401°	4′ 14″
0.2	"	=	11°	27′ 33″
0.03	"	=	1°	43′ 8″
7.23	"	=	414°	14′ 55″

PRACTICE EXERCISES FOR SECTION 15

For answers to all practice exercise problems or questions
see Section 20

1. In what respect does a foot-pound differ from a pound?
2. If 100 pounds is dropped, how much energy will it be capable of exerting after falling 10 feet?
3. Can the force of a hammer-blow be expressed in pounds?
4. If a 2-pound hammer is moving 30 feet per second, what is its kinetic energy?
5. If the hammer referred to in question 4 drives a nail into a board $\frac{1}{4}$ inch, what is the average force of the blow?
6. What relationship is there between the muzzle velocity of a projectile fired upward and the velocity with which the projectile strikes the ground?
7. What is the difference between the composition of forces and the resolution of forces?
8. If four equal forces act along lines 90 degrees apart through a given point, what is the shape of the corresponding polygon of forces?
9. Skids are to be employed for transferring boxed machinery from one floor to the floor above. If these skids are inclined at an angle of 35 degrees, what force in pounds, applied parallel to the skids, will be required to slide a boxed machine weighing 2500 pounds up the incline, assuming that the coefficient of friction is 0.20?
10. Referring to question 9, if the force or pull were applied in a horizontal direction instead of in line with the skids, what increase, if any, would be required?
11. Will the boxed machine referred to in question 9 slide down the skids by gravity?
12. At what angle will the boxed machine referred to in question 9 begin to slide by gravity?
13. What name is applied to the angle which marks the dividing line between sliding and non-sliding when a body is placed on an inclined plane?

14. How is the "angle of repose" determined?

15. What figure or value is commonly used in engineering calculations for acceleration due to gravity?

16. Is the value commonly used for acceleration due to gravity strictly accurate for any locality?

17. A flywheel 3 feet in diameter has a rim speed of 1200 feet per minute, and another flywheel 6 feet in diameter has the same rim speed. Will the rim stress or the force tending to burst the larger flywheel be greater than the force in the rim of the smaller flywheel?

18. What is the relation between the centrifugal force developed by a revolving flywheel and the force which tends to disrupt or break the flywheel rim?

19. Does the stress in the rim of a flywheel increase in proportion to the rim velocity?

20. What is generally considered the maximum safe speed for the rim of a solid or one-piece cast-iron flywheel?

21. Why is a well-constructed wooden flywheel adapted to higher speeds than one made of cast iron?

22. What is the meaning of the term "critical speed" as applied to a rotating body?

23. How is angular velocity generally expressed?

24. What is a radian and how is its angle indicated?

25. How many degrees are there in 2.82 radians?

26. How many degrees are in the following radians: $\dfrac{\pi}{3}$; $\dfrac{2\pi}{5}$; 2π?

27. Reduce to radians: 63°; 45° 32′; 6° 37′ 46″; 22° 22′ 22″.

28. Find the angular velocity of the following: 157 r.p.m.; 275 r.p.m.; 324 r.p.m.

29. Why do the values in the l column on Handbook pages 72 and 73 equal those in the radian column on page 275?

30. If the length of the arc of a sector is $4\frac{7}{8}$ inches and the radius is $6\frac{7}{8}$ inches, find the central angle.

31. A 12-inch grinding wheel has a surface speed of a mile a minute. Find its angular velocity and its revolutions per minute.

32. The radius of a circle is $1\frac{1}{2}$ inches and the central angle is 60 degrees. Find the length of the arc.

33. If an angle of 34° 12′ subtends an arc of 16.25 inches, find the radius of the arc.

SECTION 16

STRENGTH OF MATERIALS

Handbook Pages 356 to 462

This section of MACHINERY'S HANDBOOK contains fundamental formulas and data for use in proportioning parts that are common to almost every type of machine or mechanical structure. In designing machine parts, factors other than strength often are essential. For example, some parts are made much larger than required for strength alone in order to resist extreme vibrations, or deflection, or wear; consequently, many machine parts cannot be designed merely by mathematical or strength calculations and their proportions should, if possible, be based upon experience or upon similar designs that have proved successful. It is evident that no engineering handbook can take into account the endless variety of requirements relating to all types of mechanical apparatus and it is necessary for the designer to determine these local requirements for himself; but, even when the strength factor is secondary due to some other requirement, the strength, especially of the more important parts, should be calculated, in many instances, merely to prove that it will be sufficient.

In designing for strength, the part is so proportioned that the maximum working stress likely to be encountered will not exceed the strength of the material by a suitable margin. This is accomplished by the use of a factor of safety. The relationship between the working stress, s_w, the strength of the material, S_m, and the factor of safety, f_s, is given by Equation (1) on page 358 of the Handbook:

$$s_w = \frac{S_m}{f_s} \tag{a}$$

The value selected for the strength of the material, S_m, depends on the type of material, whether failure is expected to occur because of tensile, compressive, or shear stress, and on whether the stresses are constant, fluctuating, or are abruptly applied as in the case of shock loading. In general the value of S_m is based on yield strength for ductile materials, ultimate strength for brittle materials, and fatigue strength for parts subject to cyclic stresses. Moreover, the value for S_m must be for the temperature at which the part operates. Values of S_m for common materials at 68° F can be obtained from the tables in MACHINERY'S HANDBOOK from pages 444 to 447. Factors from the table given on page 446 "Influence of Temperature on the Strength of Metals" can be used to convert strength values at 68° F to values applicable at elevated temperatures. For heat treated carbon and alloy steel parts, see data on pages 2033 and 2034.

The factor of safety depends on the relative importance of reliability, weight, and cost. General recommendations are given in the Handbook on pages 358 and 359.

Working stress is dependent on the shape of the part, hence on a stress concentration factor, and on a nominal stress associated with the way in which the part is loaded. Equations and data for calculating nominal stresses, stress concentration factors, and working stresses are given on pages 358 to 370.

Example 1: — Determine the allowable working stress for a part that is to be made from SAE 1112 free cutting steel; the part is loaded in such a way that failure is expected to occur in tension when the yield strength has been exceeded. A factor of safety of 3 is to be used.

From the table, "Strength Data for Iron and Steel" on page 444 of the Handbook, a value of 30,000 psi is selected for the strength of the material, S_m. Working stress s_w is calculated from Equation (a) as follows:

$$s_w = \frac{30,000}{3} = 10,000 \text{ psi}$$

Finding Diameter of Bar to Resist Safely a Given Load. — Assume that a direct tension load, F, is applied to a bar such that the force acts along the longitudinal axis of the bar. From page 365 of the Handbook, the following equation is given for calculating the nominal stress:

$$\sigma = \frac{F}{A} \qquad\qquad (b)$$

where A is the cross-sectional area of the bar. Equation (2) on page 359 relates the nominal stress to the stress concentration factor, K, and working stress, s_w:

$$s_w = K\sigma \qquad \text{(c)}$$

Combining Equations (a), (b), and (c), there results the following:

$$\frac{S_m}{Kf_s} = \frac{F}{A} \qquad \text{(d)}$$

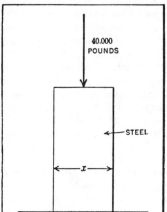

40.000 POUNDS

STEEL

x

Fig. 1. Calculating Diameter x to Safely Support a Given Load

Example 2: — A structural steel bar supports in tension a load of 40,000 pounds. The load is gradually applied, and then after having reached its maximum value, is gradually removed. Find the diameter of round bar required.

According to the table on Handbook page 444, the yield strength of structural steel is 33,000 psi. Suppose that a factor of safety of 3 is used and a stress concentration factor of 1.1. Then, inserting known values in Equation (d):

$$\frac{33,000}{1.1 \times 3} = \frac{40,000}{A} \;;\; A = \frac{40,000 \times 3.3}{33,000} \;;\; A = 4 \text{ square inches}$$

Hence, the cross-section of the bar must be about 4 square inches. As the bar is circular in section, the diameter must then be about $2\frac{1}{4}$ inches.

Diameter of Bar to Resist Compression. — If a short bar is subjected to compression in such a way that the line of application of the load coincides with the longitudinal axis of the bar, the formula for nominal stress is the same as for direct tension loading; Equation (b) and hence Equation (d) may be also applied to direct compression loading.

Example 3: — A short structural steel bar supports in compression a load of 40,000 pounds. (See Fig. 1.) The load is steady. Find the diameter of the bar required.

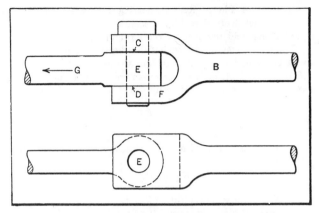

Fig. 2.   Finding the Diameter of Connecting-rod Pin to Resist
a Known Load G

From page 444 in the Handbook, the yield strength of structural steel is 33,000 psi. If a stress concentration factor of 1.1 and a factor of safety of 2.5 are used, then substituting values into Equation (d):

$$\frac{33,000}{1.1 \times 2.5} = \frac{40,000}{A} \; ; A = 3.33 \text{ square inches}$$

The diameter of a bar, the cross-section of which is 3.33 square inches is $2\frac{1}{16}$ inches approximately. (See page 55).

According to a general rule, the simple formulas that apply to compression should be used only if the length of the member being compressed is not greater than 6 times the least cross-sectional dimension. For example, these formulas should be applied to round bars only when the length of the bar is less than 6 times the diameter. If the bar is rectangular in shape, they should be applied only to bars that have a length less than 6 times the shortest side of the rectangle. When bars are longer than this, a compressive stress causes a sidewise bending action, and an even distribution of the compression stresses over the total area of the cross-section is no longer to be depended upon. Special formulas for long bars or columns will be found on Handbook page 430; see also text beginning on page 427, "Strength of Columns or Struts."

Diameter of Pin to Resist Shearing Stress. — The pin E shown in the illustration, Fig. 2, is subjected to shear. Parts G and B are held together by the pin and tend to shear it off at C and D. The areas resisting the shearing action are equal to the cross-sectional areas of the pin at these points.

From the "Table of Simple Stresses" on page 365 of MACHIN-ERY'S HANDBOOK, the equation for direct shear is:

$$\tau = \frac{F}{A} \qquad (e)$$

τ is a simple stress related to the working stress, s_w, by Equation (3) on page 359:

$$s_w = K\tau \qquad (f)$$

where K is a stress concentration factor. Combining Equations (a), (e), and (f) gives Equation (d). In this case S_m is of course the shearing strength of the material.

If a pin is subjected to shear as in Fig. 2, so that two surfaces, as at C and D, must fail by shearing before breakage occurs, the areas of both surfaces must be taken into consideration when calculating the strength. The pin is then said to be in *double shear*. If the lower part F of connecting-rod B were removed, so that member G were connected with B by a pin subjected to shear at C only, the pin would be said to be in *single shear*.

Example 4: — Assume that in Fig. 2 the load at G pulling on the connecting-rod is 20,000 pounds. The material of the pin is SAE 1025 steel. The load is applied in such a manner that shocks are liable to occur. Find the required dimensions for the pin.

Since the pins are subject to shock loading, the nominal stress resulting from the application of the 20,000 pound load must be assumed to be twice as great (see Handbook page 448) as it would be if the load were gradually applied or steady. From page 444 in the Handbook, the ultimate strength in shear for SAE 1025 steel is 75 per cent of 60,000 or 45,000 psi. A factor of safety of 3 and a stress concentration factor of 1.8 are to be used. Substituting values into Equation (d):

$$\frac{45,000}{1.8 \times 3} = \frac{2 \times 20,000}{A} \; ; A = \frac{10.8 \times 20,000}{45,000}$$
$$A = 4.8 \text{ sq. ins.}$$

As the pin is in double shear, that is, as there are two surfaces C and D over which the shearing stress is distributed, each of them must have an area of one-half the total shearing area A. In this case, then, the cross-sectional area of the pin will be 2.4 square inches, and the diameter of the pin, to give a cross-sectional area of 2.4 square inches, must be $1\frac{3}{4}$ inches.

Beams and Stresses to which they are Subjected. — Parts of machines and structures subjected to bending are known mechanically as *beams*. Hence, in this sense, a lever fixed at one end and subjected to a force at its other end, a rod supported at both ends and subjected to a load at its center, or the overhanging arm of a jib crane, would all be known as beams.

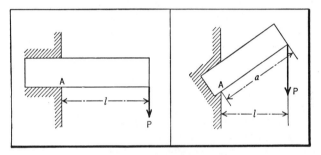

Figs. 3 and 4. Diagrams Illustrating Principle of Bending Moments

The stresses in a beam are principally tension and compression stresses. If a beam is supported at the ends and a load rests upon the upper side, the lower fibers will be stretched by the bending action and will be subjected to a tensile stress, while the upper fibers will be compressed and be subjected to a compressive stress. There will be a slight lengthening of the fibers in the lower part of the beam, while those on the upper side will be somewhat shorter, depending upon the amount of deflection. If we assume that the beam is either round or square in cross-section, there will be a layer or surface through its center line which will be neither in compression nor in tension. This surface is known as the neutral surface. The stresses of the individual layers or fibers of the beam will be proportional to their distances from the neutral surface, the stresses being greater the farther away from the neutral surface the fiber is located. Hence, there is no stress on the fibers in the neutral surface, but there is a maximum tension on the fibers at the extreme lower side and a maximum compression on the fibers at the extreme upper side of the beam. In calculating the strength of beams, it is, therefore, only necessary to determine that the fibers of the beam which are at the greatest distance from the neutral surface are not stressed beyond the safe working stress of the material. If this is the case, all the other parts of the section of the beam are not stressed beyond the safe working stress of the material.

In addition to the tension and compression stresses, a loaded beam is also subjected to a stress which tends to shear it. This shearing stress depends upon the magnitude and kind of load. In most cases, the shearing action can be ignored for metal beams, especially if the beams are long and the loads far from the supports. If the beams are very short and the load quite close to a support, then the shearing stress may become equal to or greater than the tension or compression stresses in the beam and in that case the beam should be calculated for shear.

Beam Formulas. — The bending action of a load upon a beam is called the *bending moment*. For example, in Fig. 3 the load P acting downward on the free end of the cantilever beam has a moment or bending action about the support at A equal to the load multiplied by its distance from the support. The bending moment is commonly expressed in inch-pounds, the load being expressed in pounds and the lever arm or distance from the support in inches. The length of the lever arm should always be measured in a direction at right angles to the direction of the load. Thus, in Fig. 4, the bending moment is not $P \times a$, but is $P \times l$, because l is measured in a direction at right angles to the direction of the load P.

The property of a beam to resist the bending action or the bending moment is called the *moment of resistance* of the beam. It is evident that the bending moment must be equal to the moment of resistance. The moment of resistance, in turn, is equal to the stress in the fiber farthest away from the neutral plane multiplied by the section modulus. The *section modulus* is a factor which depends upon the shape and size of the cross-section of a beam, and is given for different cross-sections in all engineering handbooks. (See table "Moments of Inertia, Section Moduli, etc., of Sections" in MACHINERY'S HANDBOOK pages 372 to 381.) The section modulus, in turn equals the *moment of inertia* of the cross-section, divided by the distance from the neutral surface to the most extreme fiber. The moment of inertia formulas for various cross-sections will also be found in the table just referred to.

The following formula on Handbook page 365 may be given as the fundamental formula for bending of beams:

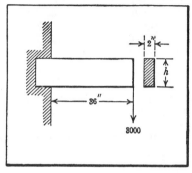

Fig. 5.   Determining the Depth *h* of a Beam to Support a Known Weight

$$\sigma = \pm \frac{M}{Z} = \pm \frac{My}{I} \qquad (g)$$

The moment of inertia I is a property of the cross-section that determines its relative strength. In calculations of strength of materials, a standard engineering handbook is necessary because of the tabulated formulas and data relating to section moduli and moments of inertia, areas of cross-sections, etc., to be found therein.

There are many different ways in which a beam can be supported and loaded, and the bending moment caused by a given load varies greatly according to whether the beam is supported at one end only or at both ends, and also whether it is freely supported at the ends or is held firmly. Then the load may be equally distributed over the full length of the beam or may be applied at one point either in the center or near to one or the other of the supports. The point where the stress is maximum is generally called the critical point. The stress at the critical point, equals bending moment divided by section modulus. Formulas for determining the stresses at the critical points will be found in the table of beam formulas, Handbook pages 404 to 415.

Example 5: — A rectangular steel bar 2 inches thick and firmly built into a wall, as shown in Fig. 5, is to support 3000 pounds at its outer end 36 inches from the wall. What would be the necessary depth *h* of the beam to support this weight safely?

The bending moment equals the load times the distance from the point of support, or, $3000 \times 36 = 108,000$ inch-pounds.

By combining Equations (a), (c), and (g), the following equation is obtained:

$$\frac{S_m}{Kf_s} = \frac{M}{Z} \tag{h}$$

If the beam is made from structural steel, the value for S_m, based on yield strength, from page 444 in the Handbook is 33,000 psi. Using a stress concentration factor of 1.1 and a factor of safety of 2.5 values may be inserted into the above equation:

$$\frac{33,000}{1.1 \times 2.5} = \frac{108,000}{Z} \; ; \; Z = \frac{2.75 \times 108,000}{33,000} \; ; \; Z = 9 \text{ inches}^3$$

The section modulus for a rectangle equals $\frac{bd^2}{6}$, in which b is the length of the shorter side and d of the longer side of the rectangle (see Handbook page 372); hence, $Z = \frac{bd^2}{6}$.

But $Z = 9$ and $b = 2$. Inserting these values into the formula, we have:

$$9 = \frac{2\,d^2}{6}$$

from which $d^2 = 27$, and $d = 5.2$ inches. This value d corresponds to dimension h in Fig. 5. Hence, the required depth of the beam to support a load of 3000 pounds at the outer end with a factor of safety of 3 would be 5.2 inches.

In calculating beams having either rectangular or circular cross-sections, the formulas on Handbook pages 418 and 419 will be found convenient to use. A beam loaded as shown by Fig. 5 is similar to the first diagram on Handbook page 418. If the formula on this page for determining height h is applied to Example 5, Fig. 5, then

$$h = \sqrt{\frac{6\,lW}{bf}} = \sqrt{\frac{6 \times 36 \times 3000}{2 \times 12,000}} = 5.2 \text{ inches}$$

In the above calculation the stress value f is equivalent to S_m/Kf_s.

Example 6: — A steel I-beam is to be used as a crane trolley track. This I-beam is to be supported at the ends and the unsupported span is 20 feet long. The maximum load is 6000 pounds and the nominal stress is not to exceed 10,000 pounds per square inch. Determine the size of standard I-beam; also the maximum deflection when the load is at the center of the beam.

The foregoing conditions are represented by Case 2, Handbook page 404. A formula for the stress at the critical point is $\frac{Wl}{4Z}$. As explained on page 403, all dimensions are in inches, and the minus sign preceding a formula merely denotes compression of the upper fibers and tension in the lower fibers.

Inserting the known values in the formula:

$$10,000 = \frac{6000 \times 240}{4\,Z}\;;\quad \text{hence}$$
$$Z = \frac{6000 \times 240}{10,000 \times 4} = 36$$

The table of standard I-beams on Handbook page 395 shows that a 12-inch I-beam which weighs 31.8 pounds per foot has a section modulus of 36.

The formula for maximum deflection (see Handbook page 405—Case 2) is $\frac{Wl^3}{48\,EI}$. According to the table on Handbook page 444, the modulus of elasticity (E) of structural steel is 29,000,000. As $Z =$ moment of inertia $I \div$ distance from neutral axis to extreme fiber (see Handbook page 403), then for a 12-inch I-beam $I = 6\,Z = 216$; hence,

$$\text{Maximum deflection} = \frac{6000 \times 240^3}{48 \times 29,000,000 \times 216} = 0.27 \text{ inch}$$

Example 7: — All conditions are the same as in Example 6, excepting that the maximum deflection at the "critical point" or center of the I-beam, must not exceed $\frac{1}{8}$ inch. What size I-beam is required?

To meet this requirement regarding deflection

$$\frac{1}{8} = \frac{Wl^3}{48\,EI}\;;\quad \text{therefore}$$
$$I = \frac{8\,Wl^3}{48\,E} = \frac{8 \times 6000 \times 240^3}{48 \times 29,000,000} = 476$$

If $x =$ distance from neutral axis to most remote fiber ($\frac{1}{2}$ beam depth in this case) then $Z = \dfrac{I}{x}$ and the table on Handbook page 395 shows that a 15-inch, 50-pound I-beam should be used because it has a section modulus of 64.2 and $\dfrac{476}{7.5} = 63.5$ nearly.

If 476 were divided by 6 ($\frac{1}{2}$ depth of a 12-inch I-beam) the result would be much higher than the section modulus of any standard 12-inch I-beam ($476 \div 6 = 79.3$); moreover, $476 \div 9 = 53$ which shows that an 18-inch I-beam is larger than is necessary because the lightest beam of this size has a section modulus of 81.9

Example 8: — If the speed of a motor is 1200 revolutions per minute and if its driving pinion has a pitch diameter of 3 inches, determine the torsional moment to which the pinion shaft is subjected, assuming that 10 horsepower is being transmitted.

If $W =$ tangential load in pounds, $H =$ the number of horsepower and $V =$ pitch-line velocity in feet per minute.

$$W = \frac{33,000 \times H}{V}$$

$$= \frac{33,000 \times 10}{943}$$

$$= 350 \text{ pounds}$$

The torsional moment $= W \times$ pitch radius of pinion $= 350 \times 1.5 = 525$ pound-inches (or inch-pounds).

Example 9: — If the pinion referred to in Example 8 drives a gear having a pitch diameter of 12 inches, to what torsional or turning moment is the gear shaft subjected?

The torque or torsional moment in any case $=$ pitch radius of gear $\times$ tangential load. The latter is the same for both gear and pinion; hence, torsional moment of gear $= 350 \times 6 = 2100$ inch-pounds.

The torsional moment or the turning effect of a force which tends to produce rotation depends (1) upon the magnitude of the force acting, and (2) upon the distance of the force from the axis of rotation, measuring along a line at right angles to the line of action of the force.

PRACTICE EXERCISES FOR SECTION 16

For answers to all practice exercise problems or questions
see Section 20

1. What is a "factor of safety" and why are different factors
   used in machine design?
2. If the ultimate strength of a steel rod is 60,000 pounds
   per square inch and the factor of safety is 5, what is
   the equivalent working stress?
3. If a steel bar must withstand a maximum pull of 9000
   pounds and if the maximum nominal stress must not
   exceed 12,000 pounds per square inch, what diameter bar
   is required?
4. Is a steel rod stronger when at ordinary room temperature
   or when heated, say, to 500° F?
5. What is the meaning of the term "elastic limit"?
6. Approximately what percentages of copper and zinc in brass
   result in the greatest tensile strength?
7. A floor beam measuring 2 by 6 inches and made of red
   spruce has an unsupported span of 10 feet. Determine the maximum safe load that could be placed in
   the center of this beam. (For working stresses of structural timbers based upon U. S. Government tests, see
   Handbook page 447.)
8. If four 10-foot long pipes are to be used to support a water
   tank installation weighing 100,000 pounds, what diameter
   standard weight pipe is required?

SECTION 17

DESIGN OF SHAFTS AND KEYS FOR POWER TRANSMISSION

Handbook pages 450 to 462

This section is a review of the general procedure in designing shafts to resist torsional and also combined torsional and bending stresses. The diameter of a shaft through which power is transmitted depends, for a given shaft material, upon the amount and kind of stress or stresses to which the shaft is subjected. In order to illustrate the general procedure, we shall assume first that the shaft is subjected only to a uniform torsional or twisting stress and that there is no additional bending stress which needs to be considered in determining the diameter.

Example 1: — A lineshaft carrying pulleys located close to the bearings is to transmit 50 horsepower at 1200 revolutions per minute. If the load is gradually applied, and steady, what diameter steel shaft is required assuming that the pulleys are fastened to the shaft by means of keys and that the bending stresses caused by the pull of the belts are negligible?

According to the former American Standard Association's Code for the Design of Transmission Shafting, the diameter of shaft required to meet the stated conditions can be determined by using the following formula (Formula 1b on page 459 of the Handbook):

$$D = B \sqrt[3]{\frac{321,000 \, K_t P}{S_s N}}$$

In this formula, D = required shaft diameter in inches; B = a factor, which for solid shafts is taken as 1; K_t = combined shock and fatigue factor; P = maximum horsepower transmitted by shaft; S_s = maximum allowable torsional shearing stress in pounds per square inch; and N = shaft speed in revolutions per minute.

From Table 1 on page 460 of the Handbook, $K_t = 1.0$ for gradually applied and steady loads, and from Table 2 the recommended maximum allowable working stress for "Commercial

Steel" shafting with keyways subjected to pure torsion loads is 6000 pounds per square inch. Substituting in the formula,

$$D = 1 \times \sqrt[3]{\frac{321,000 \times 1.0 \times 50}{6000 \times 1200}} = 1.306 \text{ inches}$$

The nearest standard size transmission shafting from the table on page 459 is $1\frac{7}{16}$ inches.

Example 2: — If in Example 1 the shaft diameter had been determined by using Formula 5b on page 451, what would the result have been and why?

$$D = \sqrt[3]{\frac{53.5P}{N}} = \sqrt[3]{\frac{53.5 \times 50}{1200}} = 1.306 \text{ inches}$$

This formula gives the same shaft diameter as was previously determined because it is a simplified form of the first formula used and contains the same values of K_t and S_s, but combined as the single constant 53.5. For lineshafts carrying pulleys under conditions ordinarily encountered, this simplified formula is usually quite satisfactory; but, where conditions of shock loading are known to exist, it is safer to use Formula 1b on page 459 which takes such conditions into account.

Shafts Subjected to Combined Stresses. — The preceding formulas are based on the assumption that the shaft is subjected to torsional stresses only. Many shafts, however, must withstand stresses that result from combinations of torsion, bending, and shock loading. In such cases it is necessary to use formulas which take such conditions into account.

Example 3: — Suppose that after the lineshaft in Example 1 was installed it became necessary to relocate a machine which was being driven by one of the pulleys on the shaft. Because of the new machine location it was necessary to move the pulley on the lineshaft further away from the nearest bearing and, as a result, a bending moment of 2000 inch-pounds has been introduced. Is the $1\frac{7}{16}$-inch diameter shaft sufficient to take this additional stress or will it be necessary to relocate the bearing to provide better support?

Since the shaft now has both bending and torsional loads acting on it, Formula 3b on page 459 should be used to compute the required shaft diameter. This diameter is then compared with the $1\frac{7}{16}$-inch diameter previously determined.

$$D = B \sqrt[3]{\frac{5.1}{p_t}} \sqrt{(K_m M)^2 + \left(\frac{63,000\ K_t P}{N}\right)^2}$$

In this formula B, K_t, P, and N are quantities previously defined and p_t = maximum allowable shearing stress under combined loading conditions in pounds per square inch; K_m = combined shock and fatigue factor; and M = maximum bending moment in inch-pounds.

From Table 1 on page 460, K_m = 1.5 for gradually applied and steady loads and from Table 2, p_t = 6000 pounds per square inch. Substituting in the formula,

$$D = 1 \times \sqrt[3]{\frac{5.1}{6000}} \sqrt{(1.5 \times 6000)^2 + \left(\frac{63,000 \times 1 \times 50}{1200}\right)^2}$$

$$= \sqrt[3]{\frac{5.1}{6000}} \sqrt{81,000,000 + 6,890,625} = \sqrt[3]{\frac{5.1}{6000} \times 9375}$$

$$= \sqrt[3]{7.968} = 1.995 \text{ inches, say, 2 inches}$$

Since this diameter is larger than the $1\frac{7}{16}$ diameter actually used for the shaft in Example 1, it will be necessary to relocate the bearing so that it is closer to the pulley, thus reducing the bending moment. This will permit the $1\frac{7}{16}$-inch diameter shaft to operate within the allowable working stress for which it was originally designed.

Design of Shafts to Resist Torsional Deflection. — In many cases shafts must be proportioned not only to provide the strength required to transmit a given torque, but also to prevent torsional deflection (twisting) through a greater angle than has been found satisfactory for a given type of service. This is particularly true in the case of machine shafts and machine-tool spindles.

For ordinary service it is customary that the angle of twist of machine shafts be limited to $\frac{1}{10}$ degree per foot of shaft length, and for machine shafts subject to load reversals, $\frac{1}{20}$ degree per foot of shaft length. As explained in the Handbook, the usual design procedure for shafting which is to have a specified maximum angular deflection is to compute the diameter of shaft required based on both deflection and strength considerations and then to choose the larger of the two diameters thus determined.

Example 4: — A 6-foot long feed shaft is to transmit a torque of 200 inch-pounds. If there are no bending stresses and the shaft is to be limited to a torsional deflection of $\frac{1}{20}$ degree per foot of length, what diameter shaft should be used? The shaft is to be made of cold drawn steel and is to be designed for a maximum working stress of 6000 pounds per square inch in torsion.

The diameter of shaft required for a maximum angular deflection α is given by Formula 8 on page 454.

$$D = 4.9 \sqrt[4]{\frac{Tl}{G\alpha}}$$

In this formula T = applied torque in inch-pounds; l = length of shaft in inches; G = torsional modulus of elasticity which for steel is 11,500,000 pounds per square inch; and α = angular deflection of shaft in degrees.

In the problem at hand, T = 200 inch-pounds; l = 6 × 12 = 72 inches; and α = 6 × $\frac{1}{20}$ = 0.3 degree.

$$D = 4.9 \sqrt[4]{\frac{200 \times 72}{11,500,000 \times 0.3}} = 4.9 \sqrt[4]{0.0041739}$$

$$= 4.9 \times 0.254 = 1.24 \text{ inches}$$

The diameter of the shaft based on strength considerations is obtained by using Formula 3a on page 451:

$$D = \sqrt[3]{\frac{5.1\,T}{S_s}} = \sqrt[3]{\frac{5.1 \times 200}{6000}} = \sqrt[3]{0.17} = 0.55 \text{ inch}$$

Therefore, since the diameter based on torsional deflection considerations is the larger of the two values obtained, the nearest standard diameter, $1\frac{1}{4}$ inches, should be used.

Selection of Key Size Based on Shaft Size. — Keys are generally proportioned in relation to shaft diameter instead of in relation to torsional load to be transmitted because of practical reasons such as standardization of keys and shafts. Thus, on Handbook page 979 will be found a table entitled "Key Size Versus Shaft Diameter." Dimensions of both square and rectangular keys are given, but for shaft diameters up to and including $6\frac{1}{2}$ inches, square keys are preferred, while for larger shafts, rectangular keys are commonly used.

Two rules which base key length on shaft size are: (1) L = 1.5 D

and (2) $L = 0.3\ D^2 \div T$, where L = length of key, D = diameter of shaft and T = key thickness.

If the keyseat is to have fillets and the key is to be chamfered, suggested dimensions for these modifications are given on page 982. If a set screw is to be used over the key, suggested sizes are given in the table on page 982.

Example 5: — If the maximum torque output of a 2-inch diameter shaft is to be transmitted to a keyed pulley, what should be the proportions of the key?

According to the table on page 979, a ½-inch square key would be preferred. If a rectangular key were selected, its dimensions would be ½ by ⅜ inch. According to rule 1 its length would be 3 inches.

The key and keyseat may be so proportioned as to provide a clearance or an interference fit. The table on page 981 gives tolerances for widths and depths of keys and keyseats to provide Class 1 (clearance) and Class 2 (interference) fits. An additional Class 3 (interference) fit which has not been standardized is mentioned on page 977 together with suggested tolerances.

Keys Proportioned According to Transmitted Torque. — As previously stated, if key sizes are based on shaft diameter there will be cases where the dimensions of the key will be excessive. This is usually true whenever a gear or pulley transmits only a portion of the total torque capacity of the shaft to which it is keyed. If excessively large keys are to be avoided, it may be advantageous in such cases to base the determination on the torque to be transmitted rather than on the shaft diameter, and to use the dimensions thus determined as a guide in selecting a standard size key.

A key proportioned to transmit a specified torque may fail in service either by shearing or by crushing, depending on the proportions of the key and the manner in which it is fitted to the shaft and hub. The best proportions for a key are those which make it equally resistant to failure by shearing and by crushing. The safe torque in inch-pounds that a key will transmit, based on the allowable shearing stress of the key material, may be found from the formula

$$T_s = L \times W \times \frac{D}{2} \times S_s \qquad (1)$$

The safe torque based on the allowable compressive stress of the key material is found from the formula

$$T_c = L \times \frac{H}{2} \times \frac{D}{2} \times S_c \qquad (2)$$

$\left(\text{For Woodruff keys the amount that the key projects above the shaft is substituted for } \dfrac{H}{2} \cdot \right)$

In these formulas, T_s = safe torque in shear; T_c = safe torque in compression; S_s = allowable shearing stress; S_c = allowable compressive stress; L = key length in inches; W = key width in inches; H = key thickness in inches; and D = shaft diameter in inches.

To satisfy the condition that the key be equally resistant to shearing and crushing, T_s should equal T_c. Thus, by equating Formulas (1) and (2), it is found that the width of the keyway in terms of the height of the keyway is

$$W = HS_c \div 2S_s \qquad (3)$$

For the type of steel commonly used in making keys, the allowable compressive stress S_c may be taken as twice the allowable shearing stress S_s of the material if the key is properly fitted on all four sides. By substituting $S_c = 2S_s$ in Formula (3) it will be found that $W = H$, so that for equal strength in compression and shear a square key should be used.

If a rectangular key is used, and the thickness H is less than the width W, then the key will be weaker in compression than in shear so that it is sufficient to check the torque capacity of the key using Formula (2).

Example 6: — A 3-inch shaft is to deliver 100 horsepower at 200 revolutions per minute through a gear keyed to the shaft. If the hub of the gear is 4 inches long, what size key, equally strong in shear and compression, should be used? The allowable compressive stress in the shaft is not to exceed 16,000 pounds per square inch and the key material has an allowable compressive stress of 20,000 pounds per square inch and an allowable shearing stress of 15,000 pounds per square inch.

The first step is to decide on the length of the key. Since the hub of the gear is 4 inches long, a key of the same length may be used. The next step is to determine the torque that the key will have to transmit. Using the formula on page 450,

$$T = \frac{63,000P}{N} = \frac{63,000 \times 100}{200} = 31,500 \text{ inch-pounds}$$

To determine the width of the key, based on the allowable shearing stress of the key material, Formula (1) is used.

$$T_s = L \times W \times \frac{D}{2} \times S_s$$

$$31,500 = 4 \times W \times \frac{3}{2} \times 15,000$$

or,

$$W = \frac{31,500 \times 2}{15,000 \times 4 \times 3} = 0.350, \text{ say, } \tfrac{3}{8} \text{ inch}$$

To determine the thickness of the key, Formula (2) is used. In using this formula, however, it should be noted that if the shaft material has a different allowable compressive stress than the key material, the lower of the two values should be used. In this case, the shaft material has the lower allowable compressive stress and the keyway in the shaft would fail by crushing before the key would. Therefore,

$$T_c = L \times \frac{H}{2} \times \frac{D}{2} \times S_c$$

$$31,500 = 4 \times \frac{H}{2} \times \frac{3}{2} \times 16,000$$

or,

$$H = \frac{31,500 \times 2 \times 2}{4 \times 3 \times 16,000} = 0.656 = \tfrac{21}{32} \text{ inch}$$

Therefore, the dimensions of the key for equal resistance to failure by shearing and crushing are $\tfrac{3}{8}$ inch wide, $\tfrac{21}{32}$ thick, and 4 inches long. If, for some reason, it is desirable to use a key shorter than 4 inches, say 2 inches, then it will be necessary to increase both the width and thickness by a factor of $4 \div 2$ if equal resistance to shearing and crushing is to be maintained. Thus, the width would be $\tfrac{3}{8} \times \tfrac{4}{2} = \tfrac{3}{4}$ inch and the thickness would be $\tfrac{21}{32} \times \tfrac{4}{2} = 1\tfrac{5}{16}$ inch for a 2-inch long key.

Set-screws Used to Transmit Torque. — For certain applications it is common practice to use set-screws to transmit torque because they are relatively inexpensive to install and permit axial adjustment of the member mounted on the shaft. However, since set-screws depend primarily on friction and the shearing force at the point of the screw, they are not especially well-suited for high torques or where sudden load changes take place.

One rule for determining the proper size of set-screw states that the diameter of the screw should equal $\frac{5}{16}$ inch plus one-eighth the shaft diameter. The holding power of set-screws selected by this rule can then be checked using the formula on page 1221 of the Handbook.

PRACTICE EXERCISES FOR SECTION 17

For answers to all practice exercise problems or questions
see Section 20

1. What is the polar section modulus of a shaft 2 inches in diameter?

2. If a 3-inch shaft is subjected to a torsional or twisting moment of 32,800 pound-inches, what is equivalent torsional or shearing stress?

3. Is the shaft referred to in Exercise 2 subjected to an excessive torsional stress?

4. If a 10-horsepower motor operating at its rated capacity connects by a belt with a 16-inch pulley on the driving shaft of a machine, what is the load tangential to the pulley rim and the resulting twisting moment on the shaft, assuming that the rim speed of the driven pulley is 600 feet per minute?

5. How is the maximum distance between bearings for steel line shafting determined?

6. What are "gib-head" keys and why are they used on some classes of work?

7. What is the distinctive feature of Woodruff keys?

8. What are the advantages of Woodruff keys?

9. If a keyseat $\frac{3}{8}$-inch wide is to be milled into a shaft $1\frac{1}{2}$-inch diameter, and if the keyseat depth is $\frac{3}{16}$ inch (as measured at one side) what is the depth from the top surface of the shaft or the amount to sink the cutter after it grazes the top of the shaft?

SECTION 18

PROBLEMS IN DESIGNING AND CUTTING GEARS

Handbook pages 730 to 976

In the design of gearing, there may be three distinct types of problems. These are: (1) determining the relative sizes of two or more gears to obtain a given speed or series of speeds; (2) determining the pitch of the gear teeth so that they will be strong enough to transmit a given amount of power; (3) calculating the dimensions of a gear of a given pitch, such as the outside diameter, the depth of the teeth, and other dimensions needed in cutting the gear.

When the term diameter is applied to a spur gear, the pitch diameter is generally referred to and not the outside diameter. In calculating the speeds of gearing, the pitch diameters are used and not the outside diameters, because when gears are in mesh the imaginary pitch circles roll in contact with each other.

Calculating Gear Speeds. — The simple rules for calculating the speeds of pulleys on Handbook page 1048 may be applied to gearing, provided either the pitch diameters of the gears or the numbers of teeth are substituted for the pulley diameters. Information on gear speeds, especially as applied to compound trains of gearing, also will be found in the section dealing with lathe change gears beginning on Handbook page 1415. When gear trains must be designed to secure unusual or fractional gear ratios, the tables of logarithms of gear ratios — Handbook pages 1420 to 1443 — will be found very useful. A practical application of these tables is shown by a number of examples beginning on Handbook page 1417.

Geared speed- or feed-changing mechanisms, especially as used on machine tools, are commonly designed to give a series of speed or feed changes in "geometrical progression." For the meaning of this term and the general formulas, see Handbook page 103. The practical application of geometrical progression in designing machine tool drives is explained on Handbook pages 689 to 693.

Example 1: — The following example illustrates the method of calculating the speed of a driven shaft in a combination belt and gear drive when the diameters of the pulleys and the pitch diameters of the gears are known, and the number of revolutions per minute of the driving shaft is given. If driving pulley *A*, Fig. 1, is 16 inches in diameter, and driven pulley *B*, 6 inches in

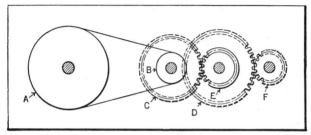

Fig. 1. Combination Pulley and Compound Gear Drive

diameter, and the pitch diameter of driving gear *C* is 12 inches, driving gear *D* is 14 inches, driven gear *E*, 7 inches, driven gear *F*, 6 inches, and driving pulley *A* makes 60 revolutions per minute, determine the number of revolutions per minute of *F*.

$$\frac{16 \times 12 \times 14}{6 \times 7 \times 6} \times 60 = 640 \text{ revolutions per minute}$$

The calculations required in solving problems of this kind can be simplified if the gears are considered as pulleys having diameters equal to their pitch diameters. When this is done, the rules that apply to compound belt drives can be used in determining the speed or size of the gears or pulleys.

If the number of teeth in each gear is substituted for its pitch diameter, the result will be the same as when the pitch diameters are used.

Example 2: — If driving spur gear *A* (Fig. 2) makes 336 revolutions per minute and has 42 teeth, driven spur gear *B*, 21 teeth, driving bevel gear *C*, 33 teeth, driven bevel gear *D*, 24 teeth, driving worm *E*, one thread, and driven worm-wheel *F*, 42 teeth, determine the number of revolutions per minute of *F*.

When a combination of spur, bevel, and worm gearing is employed to transmit motion and power from one shaft to another, the speed of the driven shaft can be found by the following method: Consider the worm as a gear having one tooth if it is single-threaded and as a gear having two teeth if double-threaded, etc. When this is done, the speed of the driving shaft can be found by applying the rules for ordinary compound spur gearing. In this example,

$$\frac{42 \times 33 \times 1}{21 \times 24 \times 42} \times 336 = 22 \text{ revolutions per minute}$$

If the pitch diameters of the gears are used instead of the number of teeth in making calculations, the worm should be considered

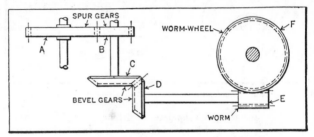

Fig. 2. Combination of Spur, Bevel, and Worm Gearing

as a gear having a pitch diameter of 1 inch, if a single-threaded, and 2 inches if a double-threaded worm, etc.

Example 3: — If a worm is triple-threaded and makes 180 revolutions per minute, and the worm-wheel is required to make 5 revolutions per minute, determine the number of teeth in the worm-wheel.

Rule: Multiply the number of threads in the worm by its number of revolutions per minute, and divide the product by the number of revolutions per minute of the worm-wheel. Applying this rule,

$$\frac{3 \times 180}{5} = 108 \text{ teeth}$$

Example 4: — A 6-inch grinding machine with a spindle speed of 1773 revolutions per minute, for a recommended peripheral speed of 6500 feet (as figured for a full-size 14-inch wheel for this size of machine), has two steps on the spindle pulley; the large step is 5.5 inches in diameter and the small step, 4 inches. What should be the minimum diameter of the wheel before the belt is shifted to the smaller step in order to obtain again a peripheral wheel speed of 6500 feet?

As the spindle makes 1773 revolutions per minute when the belt is on the large pulley, its speed with the belt on the smaller may be determined as follows:

$$5.5 : 4 = x : 1773, \text{ or } \frac{5.5 \times 1773}{4} = 2438 \text{ revolutions per min-}$$

ute, approximately. To obtain the same peripheral speed as when the belt is on the large pulley, the diameters of the grinding wheel should be $14 : x = 2438 : 1773$, or $\frac{14 \times 1773}{2438} = 10.18$ inches. Therefore, when the grinding wheel has been worn down to a diameter of 10.18 inches, or approximately $10\frac{3}{16}$ inches, the spindle belt should be shifted to the smaller step of the spindle pulley to obtain a peripheral speed of 6500 feet per minute. The method used in this example may be reduced to a formula for use with any make of grinding machine having a two-step spindle pulley.

Let D = diameter of wheel, full size;

D_1 = diameter of wheel, reduced size;

d = diameter of large pulley step;

d_1 = diameter of small pulley step;

V = spindle R.P.M., using large pulley step;

v = spindle R.P.M., using small pulley step.

Then $$v = \frac{dV}{d_1}; \qquad D_1 = \frac{DV}{v}$$

Example 5:—A geared speed-changing mechanism is to be designed to vary the speeds from 10 revolutions per minute to 200 revolutions per minute. The desired increase between successive speeds is to be about 40 per cent (see paragraph on Handbook page 690. "Ratio of Speed Changes for Machine Tools"). What number of speed changes will be required?

The geometrical ratio in this case is 1.40. See "Table of Geometrical Progression" on page 692; therefore, according to the formula on Handbook page 689,

$$1.40 = \sqrt[n-1]{\frac{200}{10}}$$

As the logarithm of a root of a number is determined by dividing the logarithm of the number by the index of the root, in this example

$$\log 1.40 = \frac{\log 20}{n-1}; \quad \text{hence}$$

$$0.14613 = \frac{1.30103}{n-1}$$

$$n - 1 = \frac{1.30103}{0.14613} = 8.9$$

$$n = 8.9 + 1 = 9.9$$

In this example, therefore, 10 changes will conform very nearly to the requirements. By means of the table on Handbook page 692 (see also instructions on page 691, under "Table of Geometrical Progression") all the speeds in a series and for any ratio listed in the first column of the table, may be determined easily. In this case, the ratio is 1.40 and the number of speeds, 10. The initial speed also is 10; hence, the second speed = 10 × 1.4 = 14; the third speed = 10 × 1.96 = 19.6, and so on up to the final speed which = 10 × 20.66 = 206.6. In actual practice, this theoretical range of speeds would be varied somewhat to permit the use of gears of standard diametral pitch.

Diametral Pitch of a Gear. — The diametral pitch represents the number of gear teeth for each inch of pitch diameter and, therefore, equals the number of teeth divided by the pitch diameter. The term diametral pitch as applied to bevel gears has the same meaning as in the case of spur gears. This method of basing the pitch on the relation between the number of teeth and the pitch diameter, is used almost exclusively in connection with cut gearing and to some extent for cast gearing. The circular pitch or the distance between the centers of adjacent teeth measured along the pitch circle is used for cast gearing, but very little for cut gearing excepting very large sizes. If 3.1416 is divided by the diametral pitch, the quotient equals the circular pitch, or if the circular pitch is known, the diametral pitch may be found by dividing 3.1416 by the circular pitch.

The pitch of gear teeth may depend primarily upon the strength required to transmit a given amount of power.

Power Transmitting Capacity of Bevel Gears. — As pointed out in the text on page 863 of the Handbook, the design of a pair of bevel gears to meet a given set of operating conditions is best accomplished in four steps: (1) Determine the design load upon which the sizes of the bevel gears will be based; (2) by means of the series of design charts given in the Handbook, select approximate gear and pinion sizes to satisfy the design load requirements; (3) check the surface durability of the gears selected in the second step by using the surface durability formula on page 866; and (4) check the strength of the gears by using the formula given on page 872. In cases where it is desired to check the capacity of an existing bevel gear drive, only steps (3) and (4) are necessary.

Dimensions and Angles Required in Producing Gears. — Many of the rules and formulas given in the gear section of MACHINERY'S HANDBOOK beginning page 730 are used in determining tooth dimensions, gear blank sizes, and also angles in the case of bevel, helical, and worm gearing. Such dimensions or angles are required on the working drawings used in connection with machining operations, such as turning gear blanks and cutting the teeth.

Example 9: — If a spur gear is to have 40 teeth of 8 diametral pitch, to what diameter should the blank be turned? Applying Formula 6a, Handbook page 746, $\frac{40 + 2}{8} = 5.25$ inches. Therefore, the outside diameter of this gear or the diameter to which the blank would be turned is $5\frac{1}{4}$ inches.

In the case of internal spur gears, the inside diameter to which the gear blank would be bored may be obtained by subtracting 2 from the number of teeth, and dividing the remainder by the diametral pitch.

Example 10: — A sample spur gear has 22 teeth, and the outside diameter, or diameter measured across the tops of the teeth, is 6 inches. Determine the diametral pitch.

According to Formula No. 6a, Handbook page 746,

$$D_O = \frac{N + 2}{P}; \quad \text{hence,}$$

$$P = \frac{N + 2}{D_O} = \frac{22 + 2}{6} = 4 \text{ diametral pitch}$$

The table, Handbook page 746, also shows that when the sample gear has American Standard stub teeth, Formula 7a should be used to determine the outside diameter, or diametral pitch.

Example 11: — A 25-degree involute full-depth spur gear is to be produced by hobbing. How is the hob tip radius found?

As shown on page 772, the maximum hob tip radius, r_c (max.), is found by the formula:

$$r_c \text{ (max.)} = \frac{0.785398 \cos \phi - b \sin \phi}{1 - \sin \phi}$$

where ϕ is the pressure angle, in this case, 25°, and b is the addendum constant which is 1.250 according to Table 1 on page 736. Thus,

$$r_c \text{ (max.)} = \frac{0.785398 \times 0.90631 - 1.25 \times 0.42262}{1 - 0.42262}$$

$$= 0.3179 \text{ inch}$$

Example 12: — If a 20-degree involute full-depth pinion having 24 teeth of 6 diametral pitch is to mesh with a rack, determine the whole depth of the rack teeth and the linear pitch of the teeth.

The teeth of a rack are of the same proportions as the teeth of a spur gear or pinion which is intended to mesh with the rack; hence the pitch of the rack teeth is equal to the circular pitch of the pinion, and is found by dividing 3.1416 by the diametral pitch.

The pitch = 3.1416 ÷ 6 = 0.5236 inch = linear pitch of rack for meshing with a pinion of 6 diametral pitch. This dimension (0.5236) represents the distance that the cutter would be indexed when milling rack teeth, or distance that the planer tool would be moved for cutting successive teeth in case the planer were used. The whole depth of a full-depth rack tooth of 20-degree pressure angle equals 2.157 divided by the diametral pitch of the meshing gear, or the whole depth equals the circular pitch multiplied by 0.6866. In this case, the circular pitch is 0.5236 and the whole depth equals 0.5236 × 0.6866 = 0.3595 inch.

Example 13: — If the teeth of a spur gear are to be cut to a certain diametral pitch, is it possible to obtain any diameter that may be desired? Thus, if the diametral pitch is 4, is it possible to make the pitch diameter $5\frac{1}{8}$ inches?

The diametral pitch system is so arranged as to provide a series of tooth sizes, just as the pitches of screw threads are standardized. Inasmuch as there must be a whole number of teeth in each gear, it is apparent that gears of a given pitch vary in diameter according to the number of teeth. Suppose, for example, that a series of gears are of 4 diametral pitch. Then the pitch diameter of a gear having, say, 20 teeth will be 5 inches; 21 teeth, $5\frac{1}{4}$ inches; 22 teeth, $5\frac{1}{2}$ inches, and so on. It will be seen that the increase in diameter for each additional tooth is equal to $\frac{1}{4}$ inch for 4 diametral pitch. Similarly for 2 diametral pitch the variations for successive numbers of teeth would equal $\frac{1}{2}$ inch, and for 10 diametral pitch the variations would equal $\frac{1}{10}$ inch, etc.

The center-to-center distance between two gears is equal to one-half the total number of teeth in the gears divided by the diametral pitch. While it may be desirable at times to have a center distance which cannot be obtained exactly by any combination of gearing of given diametral pitch, this is an unusual condition and ordinarily the designer of a machine can alter the center distance whatever slight amount may be required for gearing of the desired ratio and pitch. By using a standard system of pitches all calculations are simplified, and it is also possible to obtain the benefits of standardization in the manufacturing of gears and gear-cutters.

Proportioning Spur Gears when Center Distance is Fixed. — If the center-to-center distance between two shafts is fixed and it is desired to use gears of a certain pitch, the number of teeth in each gear for a given speed may be determined as follows: Since the gears must be of a certain pitch, the total number of teeth available should be determined and then the number of teeth in the driving and the driven gears. The total number of teeth equals twice the product of the center distance multiplied by the diametral pitch. If the center distance is 6 inches and the diametral pitch 10, the total number of teeth equals $6 \times 2 \times 10 = 120$ teeth. The next step is to find the number of teeth in the driving and the driven gears for a given rate of speed.

Rule: Divide the speed of the driving gear in revolutions per minute by the speed of the driven gear and add one to the quotient. Next divide the total number of teeth in both gears by the sum previously obtained, and the quotient will equal the number of teeth in the driving gear. This number subtracted from the total number of teeth will equal the number of teeth required in the driven gear.

Example 14: — If the center-to-center distance is 6 inches and the diametral pitch is 10, the total number of teeth available will be 120. If the speeds of the driving and the driven gears are to be 100 and 60 revolutions per minute, respectively, find the number of teeth for each gear.

$$\frac{100}{60} = 1\tfrac{2}{3} \quad \text{and} \quad 1\tfrac{2}{3} + 1 = 2\tfrac{2}{3}$$

$$120 \div 2\tfrac{2}{3} = \frac{120}{1} \times \tfrac{3}{8} = 45 = \text{number of teeth in driving gear}$$

The number of teeth in the driven gear equals 120 − 45 = 75 teeth.

When the center distance and velocity ratios are fixed by some essential construction of a machine, it is often impossible to use standard diametral pitch gear teeth. If cast gears are to be used, it does not matter so much, as a patternmaker can lay out the teeth according to the pitch desired, but if cut gears are required, an effort should be made to alter the center distance so that standard diametral pitch cutters can be used since these are usually carried in stock.

Dimensions of Generated Bevel Gears. — *Example 15:* — Find all of the dimensions and angles necessary to manufacture a pair of straight bevel gears if the number of teeth in the pinion is 16, the number of teeth in the mating gear is 49, the diametral pitch is 5, and the face width is 1.5 inches. The gears have a 20-degree pressure angle, a 90-degree shaft angle, and are to be in accordance with the Gleason System.

On Handbook page 854 is a table of formulas for Gleason System 20-degree pressure angle straight bevel gears with 90-degree shaft angle. These formulas are given in the same order as is normally used in computation. Computations of the gear dimensions should be arranged as follows for neatness and to establish a consistent procedure when calculations for bevel gears are frequently required.

Number of pinion teeth	= 16	(1)
Number of gear teeth	= 49	(2)
Diametral pitch	= 5	(3)
Face width	= 1.5	(4)
Pressure angle	= 20°	(5)
Shaft angle	= 90°	(6)

	PINION	GEAR	
Working depth	$\dfrac{2.000}{5} = 0.400$	Same as pinion	(7)
Whole depth	$\dfrac{2.188}{5} + 0.002 =$ 0.440	Same as pinion	(8)
Pitch diameter	$\tfrac{16}{5} = 3.2000$	$\tfrac{49}{5} = 9.8000$	(9)
Pitch angle	$\tan^{-1}\tfrac{16}{49} = 18°\ 5'$ *Note:* $\tan^{-1}\tfrac{16}{49}$ should be read as "the angle whose tangent is $16 \div 49$"	$90° - 18°\ 5' = 71°\ 55'$	(10)
Cone distance	$\dfrac{9.8000}{2 \times \sin 71°\ 55'} =$ 5.1546	Same as pinion	(11)
Circular pitch	$\dfrac{3.1416}{5} = 0.6283$	Same as pinion	(12)
Addendum	$0.400 - 0.116 = 0.284$	$\dfrac{0.580}{5} = 0.116$	(13)
Dedendum	$\dfrac{2.188}{5} - 0.284 =$ 0.1536	$\dfrac{2.188}{5} - 0.116 =$ 0.3216	(14)
Clearance	$0.440 - 0.400 = 0.040$	Same as pinion	(15)
Dedendum angle	$\tan^{-1}\dfrac{0.1536}{5.1546} = 1°\ 42'$	$\tan^{-1}\dfrac{0.3216}{5.1546} = 3°\ 34'$	(16)
Face angle of blank	$18°\ 5' + 3°\ 34' =$ 21° 39'	$71°\ 55' + 1°\ 42' =$ 73° 37'	(17)
Root angle	$18°\ 5' - 1°\ 42' =$ 16° 23'	$71°\ 55' - 3°\ 34' =$ 68° 21'	(18)
Outside diameter	$3.2000 + 2 \times$ $0.284 \cos 18°\ 5' =$ 3.740	$9.8000 + 2 \times$ $0.116 \cos 71°\ 55' =$ 9.872	(19)
Pitch apex to crown	$\dfrac{9.8000}{2} -$ $0.284 \sin 18°\ 5' =$ 4.812	$\dfrac{3.2000}{2} -$ $0.116 \sin 71°\ 55' =$ 1.490	(20)

	PINION	GEAR
Circular thickness	$0.6283 - 0.2530 =$ 0.3753	$\dfrac{0.6283}{2} -$ $(0.284 - 0.116)\tan 20° =$ 0.2530 (21)
Backlash	0.006	(22)
Chordal thickness	$0.3753 -$ $\dfrac{(0.3753)^3}{6 \times (3.2000)^2} -$ $\dfrac{0.006}{2} = 0.371$	$0.2530 -$ $\dfrac{(0.2530)^3}{6 \times (9.8000)^2} -$ $\dfrac{0.006}{2} = 0.250$ (23)
Chordal addendum	$0.284 +$ $\dfrac{0.3753^2 \cos 18° 5'}{4 \times 3.2000} =$ 0.294	$0.116 +$ $\dfrac{0.2530^2 \cos 71° 55'}{4 \times 9.8000} =$ 0.117 (24)

The tooth angle (Item 25) is a machine setting and is only computed if a Gleason two-tool type straight bevel gear generator is to be used. For a further description of tooth angle and Items 26 and 27, see Handbook page 853.

Dimensions of Milled Bevel Gears. — As explained on Handbook page 876, the tooth proportions of milled bevel gears differ in some respects from those of generated bevel gears. To take these differences into account a separate table of formulas is given on page 877 for use in calculating dimensions of milled bevel gears.

Example 16: — Compute the dimensions and angles of a pair of mating bevel gears that are to be cut on a milling machine using rotary formed milling cutters if the data given is as follows:

> Number of pinion teeth = 15
> Number of gear teeth = 60
> Diametral pitch = 3
> Pressure angle = $14\frac{1}{2}°$
> Shaft angle = 90°

Using the formulas on page 877,

$$\tan \alpha_P = 15 \div 60 = 0.25000 = \tan 14° 2.2', \text{ say, } 14° 2'$$
$$\alpha_G = 90° - 14° 2' = 75° 57.8', \text{ say, } 75° 58'$$
$$D_P = 15 \div 3 = 5.000 \text{ inches}$$
$$D_G = 60 \div 3 = 20.000 \text{ inches}$$

$$S = 1 \div 3 = 0.3333 \text{ inch}$$
$$S + A = 1.157 \div 3 = 0.3857 \text{ inch}$$
$$W = 2.157 \div 3 = 0.7190 \text{ inch}$$
$$T = 1.571 \div 3 = 0.5236 \text{ inch}$$
$$C = \frac{5.000}{2 \times 0.24249} = 10.308 \text{ inch}$$

(In determining C, the sine of the unrounded value of α_P, **14° 2.2′**, is used.)

$$F = 8 \div 3 = 2\tfrac{2}{3}, \text{ say, } 2\tfrac{5}{8} \text{ inches}$$
$$s = 0.3333 \times \frac{10.308 - 2\tfrac{5}{8}}{10.308} = 0.2484 \text{ inch}$$
$$t = 0.5236 \times \frac{10.308 - 2\tfrac{5}{8}}{10.308} = 0.3903 \text{ inch}$$
$$\tan \theta = 0.3333 \div 10.308 = \tan 1° 51′$$
$$\tan \phi = 0.3857 \div 10.308 = \tan 2° 9′$$
$$\gamma_P = 14° 2′ + 1° 51′ = 15° 53′$$
$$\gamma_G = 75° 58′ + 1° 51′ = 77° 49′$$
$$\delta_P = 90° - 15° 53′ = 74° 7′$$
$$\delta_G = 90° - 77° 49′ = 12° 11′$$
$$\zeta_P = 14° 2′ - 2° 9′ = 11° 53′$$
$$\zeta_G = 75° 58′ - 2° 9′ = 73° 49′$$
$$K_P = 0.3333 \times 0.97015 = 0.3234 \text{ inch}$$
$$K_G = 0.3333 \times 0.24249 = 0.0808 \text{ inch}$$
$$O_P = 5.000 + 2 \times 0.3234 = 5.6468 \text{ inches}$$
$$O_G = 20.000 + 2 \times 0.0808 = 20.1616 \text{ inches}$$
$$J_P = \frac{5.6468}{2} \times 3.5144 = 9.9226 \text{ inches}$$
$$J_G = \frac{20.1616}{2} \times 0.21590 = 2.1764 \text{ inches}$$
$$j_P = 9.9226 \times \frac{10.3097 - 2\tfrac{5}{8}}{10.3097} = 7.3961 \text{ inches}$$
$$j_G = 2.1764 \times \frac{10.3097 - 2\tfrac{5}{8}}{10.3097} = 1.6222 \text{ inches}$$
$$N'_P = \frac{15}{0.97015} = 15.4, \text{ say, } 15 \text{ teeth}$$
$$N'_G = \frac{60}{0.24249} = 247 \text{ teeth}$$

If these gears are to have uniform clearance at the bottom of the teeth, in accordance with the recommendation given in the last paragraph on page 876, then the cutting angles ζ_P and ζ_G

should be determined by subtracting the addendum angle from
the pitch cone angles. Thus,

$$\zeta_P = 14° \ 2' - 1° \ 51' = 12° \ 11'$$

$$\zeta_a = 75° \ 58' - 1° \ 51' = 74° \ 7'$$

Selection of Formed Cutters for Bevel Gears. — *Example 17:* —
In Example 16, the numbers of teeth for which to select the cutters
were calculated as 15 and 247 for the pinion and gear respectively.
Therefore, as explained on page 880 of the Handbook, the cutters
selected from the table on page 763 are the No. 7 and the No. 1
cutters. As further noted on page 880, bevel gear milling
cutters may be selected directly from the table beginning on page
878, when the shaft angle is 90 degrees, instead of using the computed value of N' to enter the table on page 763. Thus, in this
case, for a 15-tooth pinion and a 60-tooth gear, the table on
page 879 shows that the numbers of the cutters to use are 1 and 7
for gear and pinion, respectively.

Pitch of Hob for Helical Gears. — *Example 18:* — A helical
gear that is to be used for connecting parallel shafts has 83 teeth,
a helix angle of 7 degrees, and a pitch diameter of 47.78 inches.
Determine the pitch of hob to use in cutting this gear.

As explained on page 917, the normal diametral pitch or the
pitch of the hob is determined as follows: The transverse diametral pitch equals $83 \div 47.78 = 1.737$. The cosine of the helix
angle of the gear (7 degrees) is 0.99255; hence the normal diametral pitch equals $1.737 \div 0.99255 = 1.75$; therefore, a hob of
$1\frac{3}{4}$ diametral pitch should be used. This hob is the same as
would be used for spur gears of $1\frac{3}{4}$ diametral pitch, and it will
cut any spur or helical gear of that pitch regardless of the number
of teeth, provided $1\frac{3}{4}$ is the diametral pitch of the spur gear and
the normal diametral pitch of the helical gear.

Determining Contact Ratio. — As pointed out on Handbook
page 771, if a smooth transfer of load is to be obtained from one
pair of teeth to the next pair of teeth as two mating gears rotate
under load, the contact ratio must be well over 1.0. Usually,
this ratio should be 1.4 or more, although in extreme cases it
may be as low as 1.15.

Example 19: — Find the contact ratio for a pair of 18-diametral
pitch, 20-degree pressure angle gears, one having 36 teeth and

the other 90 teeth. From formula (1) given on Handbook page 770:

$$\cos A = \frac{90 \times \cos 20°}{5.111 \times 18}$$

$$= \frac{90 \times 0.93969}{91.9998}$$

$$\cos A = 0.91926 \text{ and } A$$
$$= 23°11'$$

From formula (4) on Handbook page 771:

$$\cos a = \frac{36 \times \cos 20°}{2.1111 \times 18}$$

$$= \frac{36 \times 0.93969}{37.9998}$$

$$\cos a = 0.89024 \text{ and } a$$
$$= 27°6'$$

From formula (5) on Handbook page 771:

$$\tan B = \tan 20° - \frac{36}{90}(\tan 27°6' - \tan 20°)$$

$$= 0.36397 - \frac{36}{90}(0.51172 - 0.36397)$$

$$\tan B = 0.36397 - 0.05910$$
$$= 0.30487$$

From formula (7a) on Handbook page 771, the contact ratio m_f is found:

$$m_f = \frac{90}{6.28318}(0.42826 - 0.30487)$$

$$= 1.77$$

which is satisfactory.

Dimensions Required when Using Enlarged Fine-pitch Pinions. — On Handbook page 768 there is a table of dimensions for enlarged fine-pitch pinions. This table shows how much the dimensions of enlarged pinions must differ from standard when the number of teeth is small, and undercutting of the teeth is to be avoided.

Example 20: — If a 10- and an 11-tooth mating pinion and

gear of 20 diametral pitch have both been enlarged to avoid undercutting of the teeth, what increase over standard center distance is required?

$$\text{The standard center distance} = \frac{n + N}{2P}$$

$$= \frac{10 + 11}{2 \times 20} = 0.5250 \text{ inch}$$

The amount by which the center distance must be increased over standard can be obtained by taking the sum of the amounts shown in the eighth column of Table 3 on page 768 and dividing this sum by the diametral pitch. Thus, in this case, the increase over standard center distance is $\frac{0.4151 + 0.3566}{20} = 0.0386$ inch.

Example 21: — At what center distance would the gears in Example 20 have to be meshed if there were to be no backlash?

Obtaining the two thicknesses of both gears at the standard pitch diameters from Table 3 on Handbook page 768 and using the formulas at the top of Handbook page 770:

$$\text{inv } \phi_1 = \text{inv } 20° + \frac{20\,(.09365 + .09152) - 3.1416}{10 + 11}$$

The involute of 20° is found on Handbook page 270 to be 0.014904, hence:

$$\text{inv } \phi_1 = 0.014904 + 0.026752$$
$$= 0.041656$$

Referring to the table on page 271:

$$\phi = 27°43'15''$$

$$C = \frac{10 + 11}{2 \times 20}$$

$$= 0.5250$$

$$C_1 = \frac{\cos 20°}{\cos 27°43'15''} \times 0.5250$$

$$= \frac{0.93969}{0.88522} \times 0.5250$$

$$C_1 = 0.5573 \text{ inch}$$

End Thrust of Helical Gears Applied to Parallel Shafts. — *Example 22:* — The diagrams on Handbook pages 920 to 922 show the application of helical or spiral gears to parallel shaft drives. If 7 horsepower is to be transmitted at a pitch-line velocity of 200 feet per minute, determine the end thrust in pounds, assuming that the helix angle of the gear is 15 degrees.

In order to determine the end thrust of helical gearing as applied to parallel shafts, first calculate the tangential load on the gear teeth.

$$\text{Tangential load} = \frac{33,000 \times 7}{200} = 1155 \text{ pounds}$$

The axial or end thrust may now be determined approximately by multiplying the tangential load by the tangent of the tooth angle. Thus, in this instance, the thrust = 1155 × tan 15 degrees = about 310 pounds. (Note that this formula agrees with the one on Handbook page 305 for determining force *P* parallel to base of inclined plane.) The end thrust obtained by this calculation will be somewhat greater than the actual end thrust, because frictional losses in the shaft bearings, etc., have not been taken into account, although a test on a helical gear set, with a motor drive, showed that the actual thrust of the $7\frac{1}{2}$-degree helical gears tested, was not much below the values calculated as just explained.

According to most text-books, the maximum angle for single helical gears should be about 20 degrees, although one prominent manufacturer mentions that the maximum angle for industrial drives ordinarily does not exceed 10 degrees, and this will give quiet running without excessive end thrust. On some of the heavier single helical gearing used for street railway transmissions, etc., an angle of 7 degrees is employed.

Dimensions of Worm-wheel Blank and the Gashing Angle. — *Example 23:* — A worm-wheel having 45 teeth is to be driven by a double threaded worm having an outside diameter of $2\frac{1}{2}$ inches and a lead of 1 inch, the linear pitch being $\frac{1}{2}$ inch. The throat diameter and throat radius of the worm-wheel are required, and also the angle for gashing the blank.

The throat diameter is found by applying rule or formula No. 28 on Handbook page 897. The addendum of the worm thread equals the linear pitch multiplied by 0.3183, and in this case

0.5 × 0.3183 = 0.1591 inch. The pitch diameter of the worm-wheel = 45 × 0.5 ÷ 3.1416 = 7.162 inches; hence, the throat diameter equals 7.162 + 2 × 0.1591 = 7.48 inches.

The radius of the worm-wheel throat is found by subtracting twice the addendum of the worm thread from ½ the outside diameter of the worm. The addendum of the worm thread equals 0.1591 inch, and the radius of the throat, therefore, equals (2.5 ÷ 2) − 2 × 0.1591 = 0.931 inch.

When a worm-wheel is hobbed in a milling machine, gashes are milled before the hobbing operation. The table must be swiveled around while gashing, the amount depending upon the relation between the lead of the worm thread and the pitch circumference. The first step is to find the circumference of the pitch circle of the worm. The pitch diameter equals the outside diameter minus twice the addendum of the worm thread; hence, the pitch diameter equals 2.5 − 2 × 0.1591 = 2.18 inches, and the pitch circumference equals 2.18 × 3.1416 = 6.848 inches.

Next divide the lead of the worm thread by the pitch circumference to obtain the tangent of the desired angle, and then refer to a table of tangents to determine what this angle is. In this case it is 1 ÷ 6.848 = 0.1460 which is the tangent of 8⅓ degrees, nearly. Therefore, the table of the milling machine is set at an angle of 8⅓ degrees from its normal position.

Change Gear Ratio for Diametral-pitch Worms. — *Example 24:* — In cutting worms to a given diametral pitch, the ratio of the change gears as given on Handbook page 907, is $\frac{22 \times \text{Threads per Inch}}{7 \times \text{Diametral Pitch}}$. Explain why the constants 22 and 7 are used.

The reason why the constants 22 and 7 are used in determining the ratio of change-gears for cutting worm threads, is because $\frac{22}{7}$ equals, very nearly, 3.1416, which is the circular pitch equivalent to 1 diametral pitch.

Assume that the diametral pitch of the worm-gear is 5, and the lathe screw constant is 4. (See Handbook page 1415 for the meaning of "lathe screw constant.") Then, $\frac{4 \times 22}{5 \times 7} = \frac{88}{35}$. If this simple combination of gearing were used, the gear on the stud would have 88 teeth, and the gear on the lead screw, 35 teeth. Of course, any other combination of gearing having this same ratio could be used, as for example, the following compound train of gearing: $\dfrac{24 \times 66}{30 \times 21}$.

If the lathe screw constant is 4, as previously assumed, then the number of threads per inch obtained with gearing having a ratio of $\dfrac{88}{35} = \dfrac{4 \times 35}{88} = 1.5909$; hence, the pitch of the worm thread equals $1 \div 1.5909 = 0.6284$ inch, which is the circular pitch equivalent to 5 diametral pitch, correct to within 0.0001 inch.

Bearing Loads Produced by Bevel Gears (Handbook page 862). — In applications where bevel gears are used, not only must the gears be proportioned with regard to the power to be transmitted, but also the bearings supporting the gear shafts must be of adequate size and design to sustain the radial and thrust loads that will be imposed on them. Assuming that suitable gear and pinion proportions have been selected, the next step is to compute the loads needed to determine whether or not adequate bearings can be provided. To find the loads on the bearings, first, use the formulas on Handbook page 863 to compute the tangential, axial, and separating components of the load on the tooth surfaces. Secondly, use the principle of moments, together with the components determined in the first step, to find the radial loads on the bearings. To illustrate the procedure, the following example will be used.

Example 25: — A 16-tooth left-hand spiral pinion rotating clockwise at 1800 rpm transmits 71 horsepower to a 49-tooth mating gear. If the pressure angle is 20 degrees, the spiral angle is 35 degrees, the face width is 1.5 inches, and the diametral pitch is 5, what are the radial and thrust loads that govern the selection of bearings?

In Fig. 3, the locations of the bearings for the gear shafts are shown. It should be noted that distances K, L, M, and N are measured from the center line of the bearings and from the midfaces of the gears at their mean pitch diameters. In this example it will be assumed that these distances are given and are as follows: $K = 2.5$ inches; $N = 3.5$ inches; $L = 1.5$ inches; and $M = 5.0$ inches.

Other quantities that will be required in the solution of this example are the pitch diameter, pitch angle, and mean pitch diameter of both the gear and pinion. These are computed using formulas given in the Handbook.

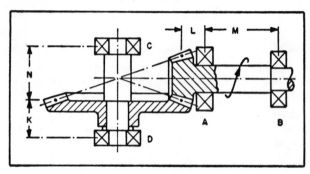

Fig. 3. Diagram Showing Location of Bearings for Bevel Gear Drive in Example 25

Using Formula 9 from page 858,

> Pitch diam. of pinion $d = 3.2$ inches
> Pitch diam. of gear $D = 9.8$ inches

Using Formula 10 from page 858,

> Pitch angle of pinion $\gamma = 18° 5'$
> Pitch angle of gear $\Gamma = 71° 55'$

Using the formula given on page 863,

> Mean pitch diameter of pinion $d_m = d - F \sin \gamma$
> $\qquad\qquad\qquad = 3.2 - 1.5 \times 0.31040$
> $\qquad\qquad\qquad = 2.734$ inches

> Mean pitch diameter of gear $D_m = D - F \sin \Gamma$
> $\qquad\qquad\qquad = 9.8 - 1.5 \times 0.95061$
> $\qquad\qquad\qquad = 8.374$ inches

The first step in determining the bearing loads is to compute the tangential, axial, and separating components of the tooth load. Using the formulas on page 863,

$$W_t = \frac{126{,}050\,P}{n d_m} = \frac{126{,}050 \times 71}{1800 \times 2.734} = 1819 \text{ pounds}$$

$$W_x \text{ for the pinion} = \frac{W_t}{\cos \psi} (\tan \phi \sin \gamma_d + \sin \psi \cos \gamma_d)$$

$$= \frac{1819}{0.81915} (0.36397 \times 0.31040 + 0.57358 \times 0.95061)$$

$$= 1462 \text{ pounds}$$

$$W_x \text{ for the gear} = \frac{W_t}{\cos \psi} (\tan \phi \sin \gamma_D - \sin \psi \cos \gamma_D)$$

$$= \frac{1819}{0.81915} (0.36397 \times 0.95061 - 0.57358 \times 0.31040)$$

$$= 373 \text{ pounds}$$

$$W_s \text{ for the pinion} = \frac{W_t}{\cos \psi} (\tan \phi \cos \gamma_d - \sin \psi \sin \gamma_d)$$

$$= \frac{1819}{0.81915} (0.36397 \times 0.95061 - 0.57358 \times 0.31040)$$

$$= 373 \text{ pounds}$$

$$W_s \text{ for the gear} = \frac{W_t}{\cos \psi} (\tan \phi \cos \gamma_D + \sin \psi \sin \gamma_D)$$

$$= \frac{1819}{0.81915} (0.36397 \times 0.31040 + 0.57358 \times 0.95061)$$

$$= 1462 \text{ pounds}$$

As explained on Handbook page 862, the axial thrust load on the bearings is equal to the axial component of the tooth load W_x. Since thrust loads are always taken up at only one mounting point, either bearing A or bearing B must be a bearing capable of taking a thrust of 1462 pounds, and either bearing C or bearing D must be capable of taking a thrust of 373 pounds.

The next step is to determine the magnitudes of the radial loads on the bearings A, B, C, and D. For an *overhung mounted* gear, or pinion, it can be shown, using the principle of moments, that the radial load on bearing A is:

$$R_A = \frac{1}{M} \sqrt{[W_t(L + M)]^2 + [W_s(L + M) - W_x r]^2} \qquad (1)$$

And the radial load on bearing B is:

$$R_B = \frac{1}{M} \sqrt{(W_t L)^2 + (W_s L - W_x r)^2} \qquad (2)$$

For a *straddle mounted* gear or pinion the radial load on bearing C is:

$$R_C = \frac{1}{N + K} \sqrt{(W_t K)^2 + (W_s K - W_x r)^2} \qquad (3)$$

And the radial load on bearing D is:

$$R_D = \frac{1}{N + K} \sqrt{(W_t N)^2 + (W_s N + W_x r)^2} \qquad (4)$$

In these formulas, r is the mean pitch radius of the gear or pinion.

These formulas will now be applied to the gear and pinion bearings in the example. Since an overhung mounting is used for the pinion, Formulas (1) and (2) are used to determine the radial loads on the pinion bearings:

$$R_A = \tfrac{1}{5}\sqrt{[1819(1.5 + 5)]^2 + [373(1.5 + 5) - 1462 \times 1.367]^2}$$
$$= 2365 \text{ pounds}$$

$$R_B = \tfrac{1}{5}\sqrt{(1819 \times 1.5)^2 + (373 \times 1.5 - 1462 \times 1.367)^2}$$
$$= 618 \text{ pounds}$$

Since a straddle mounting is used for the gear, Formulas (3) and (4) are used to determine the radial loads on the gear bearings:

$$R_C = \frac{1}{3.5 + 2.5}\sqrt{(1819 \times 2.5)^2 + (1462 \times 2.5 - 373 \times 4.187)^2}$$
$$= 833 \text{ pounds}$$

$$R_D = \frac{1}{3.5 + 2.5}\sqrt{(1819 \times 3.5)^2 + (1462 \times 3.5 + 373 \times 4.187)^2}$$
$$= 1533 \text{ pounds}$$

These radial loads, and the thrust loads previously computed, are then used to select suitable bearings from manufacturers' catalogs.

It should be noted, in applying Formulas (1) to (4), that if both gear and pinion had overhung mountings, then Formulas (1) and (2) would have been used for both; if both gear and pinion had straddle mountings, then Formulas (3) and (4) would have been used for both. In any case, the dimensions and loads for the corresponding member must be used. Also, in applying the formulas, the computed values of W_x and W_s, if they are negative, must be used in accordance with the rules applicable to negative numbers.

PRACTICE EXERCISES FOR SECTION 18

For answers to all practice exercise problems or questions see
Section 20

1. A spur gear of 6 diametral pitch has an outside diameter of 3.3333 inches. How many teeth has it? What is the pitch diameter? What is the tooth thickness measured along the pitch circle?

2. A gear of 6 diametral pitch has 14 teeth. Find the outside diameter, the pitch diameter, and the addendum.

3. When is the 25-degree tooth form standard preferred?

4. What dimension does a gear-tooth vernier caliper measure?

5. What are the principal 20-degree pressure angle tooth dimensions for the following diametral pitches: 4; 6; 8; 18?

6. Give the important $14\frac{1}{2}$-degree pressure angle tooth dimensions for the following circular pitches: $\frac{1}{2}$ inch; $\frac{3}{4}$ inch; $\frac{9}{16}$ inch.

7. What two principal factors are taken into consideration in determining the power transmitting capacity of spur gears?

8. The table on Handbook page 763 shows that a No. 8 formed cutter (involute system) would be used for milling either a 12- or 13-tooth pinion, whereas a No. 7 would be used for tooth numbers from 14 to 16, inclusive. If the pitch is not changed, why is it necessary to use different cutter numbers?

9. Are hobs made in series or numbers for each pitch similar to formed cutters?

10. If the teeth of a gear have a $\frac{8}{8}$ pitch, what name is applied to the tooth form?

11. A stub-tooth gear has $\frac{8}{10}$ pitch. What do the figures 8 and 10 indicate?

12. What is the module of a gear?

13. Explain the use of the table of chordal thicknesses on Handbook page 761.

14. Give the dimensions of a 20-degree stub tooth of 12 pitch.

15. What are the recommended diametral pitches for fine-pitch standard gears?

16. What tooth numbers could be used in pairs of gears having the following ratios: 0.2642; 0.9615?

17. What amount of backlash is provided for general purpose gearing and how is the excess depth of cut to obtain it calculated?

18. What diametral pitches correspond to the following modules: 2.75; 4; 8?

19. What is the general rule for determining the face width of bevel gears?

20. Can bevel gears be cut by formed milling cutters?

21. Can the formed cutters used for cutting spur gears also be used for bevel gears?

22. What is the minimum number of teeth recommended for bevel gears of 2 to 1 ratio, assuming that the gears have straight (not spiral) teeth?

23. For spiral bevel gears of 20-degree pressure angle, is the minimum number of teeth recommended for a 2 to 1 ratio smaller than that recommended for straight bevel gears in answer to question 22?

24. What is the pitch angle of a bevel gear?

25. Why is the face cone of a generated bevel gear made parallel to the root cone of its mating gear?

26. When is the term "miter" applied to bevel gears?

27. What is the difference between the terms "whole depth" and "working depth" as applied to gear teeth?

28. Why do pre-shaved gears have a greater dedendum than gears that are finish-hobbed?

29. What is an odontograph table?

30. Are gear teeth of 8 diametral pitch larger or smaller than teeth of 4 diametral pitch, and how do these two pitches compare in regard to tooth depth and thickness?

31. Where is the pitch diameter of a bevel gear measured?

32. What is the relation between the circular pitch of a worm gear and the linear pitch of the mating worm?

33. From what diameter is the helix angle of a worm calculated?

34. In what respect does the helix angle of a worm differ from the helix angle of a helical or spiral gear?

35. How do the terms "pitch" and "lead," as applied to a worm, compare with the same terms as applied to screw threads?

36. Find the helix angles of the following worms: Lead $\frac{3}{4}$ inch and pitch diameter 2 inches; lead $1\frac{7}{8}$ inches and pitch diameter 4 inches.

37. Find the lead of the following worms: Helix angle 19 degrees and pitch diameter 3 inches; helix angle 6 degrees and pitch diameter $3\frac{7}{8}$ inches.

38. Under what conditions is worm gearing self-locking or incapable of being reversed by driving from worm gear to worm?

39. Why is the outside diameter of a hob for cutting a worm-wheel somewhat larger than the outside diameter of the worm?

40. How is the number of flutes in a worm-gear hob determined?

41. Why are triple, quadruple, or other multiple-threaded worms used when an efficient transmission is required?

42. In designing worm drives having multi-threaded worms, it is common practice to select a number of wheel teeth that is not an exact multiple of the number of worm threads. Why is this done? When should this practice be avoided?

43. Explain the following terms used in connection with helical or spiral gears: Transverse diametral pitch; normal diametral pitch. What is the relation between these terms?

44. Are helical gear calculations based upon diametral pitch or circular pitch?

45. Can helical gears be cut with the formed cutters used for spur gears?

46. In spiral gearing the tangent of the tooth or helix angle = the circumference ÷ lead. Is this circumference calculated from the outside diameter, the pitch diameter, or the root diameter?

47. What advantages are claimed for gearing of the herring-bone type?

SECTION 19

GENERAL REVIEW QUESTIONS

For answers, see Section 20

1. One of the steels in the S. A. E. Specifications is S. A. E. nickel steel 2330. What does the number signify?

2. What does the number of a Jarno taper indicate?

3. What is the general rule for determining the direction in which to apply tolerances?

4. Why is 1 horsepower equivalent to 33,000 foot-pounds of work per minute? Why not 30,000 or some other number?

5. What is the chief element in the composition of babbitt metals?

6. If the pitch of a stub tooth gear is $\frac{8}{10}$, what is the tooth depth?

7. What does the figure 8 mean if the pitch of a stub tooth gear is $\frac{8}{10}$?

8. Explain how to determine the diametral pitch of a spur gear from a sample gear.

9. If a sample gear is cut to circular pitch, how can this pitch be determined?

10. What gage is used for seamless tubing and does it apply to all metals?

11. How does the strength of iron wire rope compare with steel rope?

12. Is the friction between two bearing surfaces proportional to the pressure?

13. If surfaces are well lubricated, upon what does frictional resistance depend?

14. What is the general rule for subtracting a negative number from a positive number? For example, $8 - (-4) = ?$

15. Is one meter longer than one yard?

16. On Handbook page 2379 two of the equivalents of one horsepower-hour are: 1,980,000 foot-pounds, and 2.64 pounds of water evaporated at 212 degrees F. How is this relationship between work and heat established?

174

17. Is "extra strong" and "double extra strong" wrought or steel pipe larger in diameter than standard weight pipe?

18. Is wrought iron pipe better than steel pipe?

19. Are the nominal sizes of wrought or steel pipe ever designated by giving the outside diameter?

20. What form of flanged fitting is known as a lateral?

21. Will charcoal ignite at a lower temperature than dry pine?

22. What general classes of steel are referred to as "stainless"?

23. What kind of steel is known as Bessemer screw stock?

24. Does the nominal length of a file include the length of the tang? For example, is a 12-inch file 12 inches long over all?

25. Is steel heavier than cast iron?

26. Is there any alloy that will melt at a temperature below the boiling point of water?

27. What is the specific gravity (a) of solid bodies, (b) of liquids, (c) of gases?

28. A system of four-digit designations for wrought aluminum and aluminum alloys was adopted by The Aluminum Association in 1954. What do the various digits signify?

29. What alloy is known as "red brass" and how does it compare with "yellow brass"?

30. What is the difference between adiabatic expansion or compression and isothermal expansion or compression?

31. Are the sizes of all small twist drills designated by numbers?

32. Why are steel tools frequently heated in molten baths for hardening them?

33. In hardening tool steel, what is the best temperature for refining the grain of the steel?

34. In cutting a screw thread on a tap assume that the pitch is to be increased from 0.125 inch to 0.1255 inch to compensate for shrinkage in hardening. How can this be done?

35. What is the general rule for reading a vernier scale (a) for linear measurements; (b) for angular measurements?

36. The end of a shaft is to be turned to a taper of $\frac{3}{4}$ inch per foot for length of 5 inches without leaving a shoulder at the end of the cut. How is the diameter of the small end determined?

37. How does the metric horsepower compare with a horse-power equivalent to 33,000 foot-pounds per minute?

38. What decimal part of a degree is 53 minutes?

39. If $10x - 5 = 3x + 16$, what is the value of x?

40. Approximately what angle is required for a cone clutch to prevent either slipping or excessive wedging action?

41. What is the coefficient of friction?

42. Is Stub's steel wire gage used for the same purpose as Stub's iron wire gage?

43. Why are some ratchet mechanisms equipped with two pawls of different lengths?

44. How should a leather belt be applied to pulleys?

45. If a quarter-turn belt drive cannot be avoided, how should the driving and driven pulleys be aligned?

46. Is the ultimate strength of a crane or hoisting chain equal to twice the ultimate strength of the bar or rod used for making the links?

47. How does the strength of a chain with studded links compare with that of an unstudded chain?

48. If a shaft $3\frac{1}{2}$ inches in diameter is to be turned at a cutting speed of 90 feet per minute, what number of revolutions per minute will be required?

49. In lapping by the "wet method," what kind of lubricant is preferable (a) with a steel lap, (b) with a cast-iron lap?

50. What is the meaning of the terms right-hand and left-hand as applied to helical or spiral gears, and how is the "hand" of the gear determined?

51. Are mating helical or spiral gears always made to the same hand?

52. How would you determine the total weight of 100 feet of $1\frac{1}{2}$-inch standard weight pipe?

53. What is the difference between casehardening and pack-hardening?

54. What is the nitriding process of heat-treating steel?

55. What is the difference between single-cut and double-cut files?

56. For general purposes what is the usual height of work benches?

57. What do the terms "major diameter" and "minor diameter" mean as applied to screw threads in connection with the American Standard?

58. Is the present S. A. E. Standard for screw threads the same as the Unified and American Standard?

59. Does the machinability of steel depend only upon its hardness or is composition also a factor?

60. Is there any direct relationship between the hardness of steel and its strength?

61. By what process is most structural steel produced?

62. Is tool steel ever made by the open-hearth process?

63. What is the recommended cutting speed in feet per minute for turning normalized AISI 4320 alloy steel with a Bhn hardness of 250, when using a carbide tool?

64. The diametral pitch of a spur gear equals the number of teeth divided by pitch diameter. Is the diametral pitch of the cutter or hob for a helical or spiral gear determined in the same way?

65. What is the difference between the Fellows and the Nuttall systems of stub gear teeth?

66. Under what conditions are gear teeth laid out by using an odontograph table?

67. How is the odontograph table used for drawing teeth that have approximately the same shape as involute teeth?

68. Is it necessary in making ordinary working drawings of gears to lay out the tooth curves by means of an odontograph table or otherwise?

69. In milling plate cams on a milling machine, how is the cam rise varied other than by changing the gears between the dividing head and feed screw?

70. How is the angle of the dividing head spindle determined for milling plate cams?

71. How is the center-to-center distance between two gears determined if the number of teeth and diametral pitch are known?

72. How is the center-to-center distance determined in the case of internal gears?

73. In the failure of riveted joints, rivets may fail through one or two cross-sections or by crushing. How may plates fail?

74. What gage is used in England to designate wire sizes?

75. What is a Prony brake?

76. What is the advantage of a Prony brake or other form of dynamometer for measuring power?

77. If a beam supported at each end is uniformly loaded throughout its length, will its load capacity exceed that of a similar beam loaded at the center only?

78. How does cypress compare with southern yellow pine in regard to strength to resist bending stresses?

79. Is the outside diameter of a 2-inch pipe about 2 inches?

80. The hub of a lever 10 inches long is secured to a 1-inch shaft by a taper pin. If the maximum pull at the end of the lever equals 60 pounds, what pin diameter is required? (Give mean diameter or diameter at center.)

81. Why are the logarithms of all numbers between 10 and 100 (not including 10 and 100) equal to 1, plus some fraction?

82. What is the logarithm (*a*) of 11; (*b*) of 99?

83. How is the pressure of water in pounds per square inch determined for any depth?

84. (*a*) What thrust is required for drilling AISI 1112 steel with a $\frac{7}{8}$-inch drill at a feed of .008 inch per revolution? (*b*) What would be the horsepower at 100 rpm?

85. If a machine producing 50 parts per day is replaced by a machine that produces 100 parts per day, what is the percentage of increase?

86. If production is decreased from 100 to 50, what is the percentage of reduction?

87. What kind of steel is used ordinarily for springs in the automotive industry?

88. What is the heat-treating process known as "normalizing"?

89. What important standards apply to electric motors?

90. Is there an American standard for section linings to represent different materials on drawings?

91. Is the taper per foot of the Morse standard uniform for all numbers or sizes?

92. Is there more than one way to remove a tap that has broken in the hole during tapping?

93. The center-to-center distance between two bearings for gears is to be 10 inches, with a tolerance of 0.005 inch. Should this tolerance be (*a*) unilateral and plus, (*b*) unilateral and minus, (*c*) bilateral?

94. How are the available pitch diameter tolerances for Acme screw threads obtained?

95. On Handbook page 1853, there is a rule for determining the pressure required for punching circular holes into steel sheets or plates. Why is the product of the hole diameter and stock thickness multiplied by 80 to obtain the approximate pressure in tons?

96. What gage is used in the United States for cold-rolled sheet steel?

97. What gage is used for brass wire, and is the same gage used for brass sheets?

98. Is the term "babbitt metal" applied to a single composition?

99. Next to tin, what are the chief elements in high-grade babbitt metal?

100. How many bars of stock 20 feet long will be needed to make 20,000 dowel-pins 2 inches long, if the tool for cutting them off is 0.100 inch wide?

101. What is the melting point and weight per cubic inch of cast iron; steel; lead; copper; nickel; tin?

102. What lubricant is recommended for machining aluminum?

103. What relief angles are recommended for cutting copper, brass, bronze and aluminum?

104. Why is stock annealed between drawing operations in producing parts in drawing dies?

105. When is it advisable to mill screw threads?

106. How does a fluted chucking reamer differ from a rose chucking reamer?

107. What kind of steel is commonly used for gages?

108. Should limit gages be marked "max size" and "min size" or is it preferable to mark them "go" and "not go"?

109. What is the "lead" of a milling machine?

110. The table on Handbook page 1464 shows that a lead of 9.625 inches will be obtained if the numbers of teeth in the *driven* gears are 44 and 28, and the numbers of teeth on the *driving* gears 32 and 40. Prove that this lead of 9.625 inches is correct.

111. Use the prime number and factor table beginning on Handbook page 86 to reduce the following fractions to their lowest terms: $\frac{210}{462}$; $\frac{2795}{6405}$; $\frac{741}{1131}$.

112. If a bevel gear and a spur gear each have 30 teeth of 4 diametral pitch, how do the tooth sizes compare?

113. For what types of work are the following machinists' files used: (a) flat files? (b) half round files? (c) hand files? (d) knife files? (e) general purpose files? (f) pillar files?

114. Referring to the illustration on Handbook page 1556, what is the dimension x over the rods used for measuring the dovetail slide if a is 4 inches, angle α is 60 degrees, and the diameter of the rods used is ⅝ inch?

115. Determine the diameter of the bar or rod for making the links of a single chain required to lift safely a load of 6 tons.

116. Why will a helical gear have a greater tendency to slip on an arbor while the teeth are being milled than when milling a straight tooth gear?

117. How does a collapsing tap reduce the time for tapping?

118. When are removable or "slip" bushings used in a jig?

119. What are the relative ratings and properties of an M30 molybdenum high speed tool steel?

120. What systematic procedure may be used in designing a roller chain drive to meet certain requirements as to horsepower, center distance, etc.?

121. Is there a precise method of indicating on a drawing the quality of a finished surface?

122. What gear steels would you use (1) for casehardened gears? (2) for fully hardened gears? (3) for gears which are to be machined after heat-treatment?

123. Is it practicable to tap holes and obtain (1) Class 2 fits? (2) Class 3 fits?

124. What is the maximum safe operating speed of an organic bonded Type 1 grinding wheel when used in a bench grinder?

125. What is the recommended type of diamond wheel and abrasive specification for internal grinding?

126. Is there a standard direction of rotation for all types of non-reversing electric motors?

127. Anti-friction bearings are normally grease lubricated. Is oil ever used? If so, when?

SECTION 20

ANSWERS TO PROBLEMS AND QUESTIONS

Section Number	Number of Question	Answers (Or where information is given in Handbook)
1	1	Table beginning page 2; also page 1
	2	Table beginning page 45; also table page 43
	3	Table beginning page 2; also table beginning page 52
	4	For powers, tables beginning pages 45 and 52; for roots, pages 14, 8, 53, 54
	5	Page 54; page 1
	6	Pages 9 and 41
	7	Table beginning page 2
	8	2.3263
	9	0.0081018
	10	0.01736
2	1	Table beginning page 55
	2	Table beginning page 67
	3	Table page 71
	4	Table page 51
	5	Table beginning page 55
	6	Table beginning page 55
	7	85.4118 square inches
	8	8 hours, 50 minutes
	9	2450.448 pounds
	10	$2\frac{1}{16}$ inches
	11	7 degrees, 10 minutes
	12	Obtain longitudinal and lateral adjustments from constants, page 78

Section Number	Number of Question	Answers (Or where information is given in Handbook)
3	1	Page 77
	2	(a) 0.043 inch (b) 0.055 inch (c) 0.102 inch
	3	0.336 inch
	4	2.796 inches
	5	4.743 inches
	6	4.221 feet
	7	Page 72
	8	740 gallons, approximately
	9	Table beginning page 82
	10	Table beginning page 82
	11	Table beginning page 82
	12	Table, page 85
4	1	(a) 104 horsepower: (b) If reciprocal is used, $H = 0.33\,D^2SN$.
	2	65 inches
	3	5.74 inches
	4	Side $s = 5.77$ inches; diagonal $d = 8.165$ inches, and volume $= 192.1$ cubic inches; $d = 8.165$ inches
	5	91.0408 square inches
	6	$0.7854 = \frac{1}{4} \times 3.1416$ or π; 3.1416 is the ratio of the circumference of a circle to its diameter; 14.7 pounds per square inch is generally assumed in engineering calculations to be atmospheric pressure at sea level; 32.16 feet per second (usually denoted by g) is the value used in engineering calculations for the acceleration due to gravity; 64.32 or 2×32.16 is a constant used, for example, in the kinetic energy formula, page 329; 144 = the number of square inches in

Section Number	Number of Question	Answers (Or where information is given in Handbook)
4		one square foot; 778 = the number of foot-pounds equivalent to one B.T.U. or British thermal unit; 1728 = number of cubic inches in one cubic foot; 33,000 = number of foot-pounds per minute equivalent to one horsepower
	7	Page 81
	8	Page 81
	9	4.1888 and 0.5236
	10	59.217 cubic inches
	11	Page 288
	12	$a = \dfrac{2\,A}{h} - b$
	13	$r = \sqrt{R^2 - \dfrac{s^2}{4}}$
	14	$a = \sqrt{\dfrac{\left(\dfrac{P}{\pi}\right)^2}{2} - b^2}$
	15	$\sin A = \sqrt{1 - \cos^2 A}$
	16	$a = \dfrac{b \times \sin A}{\sin B};\quad b = \dfrac{a \times \sin B}{\sin A};$ $\sin A = \dfrac{a \times \sin B}{b};\quad \sin B = \dfrac{b \times \sin A}{a}$
5	1	Page 120
	2	Table beginning page 126
	3	2; 2; 1; $\bar{3}$; 3; 1
	4	As location of decimal point is indicated by characteristic which is not given, the numbers might be 7082, 708.2, 70.82, 7.082, 0.7082, 0.07082, etc.; 7675, 767.5, etc.; 1689, 168.9, etc.

Section Number	Number of Question	Answers (Or where information is given in Handbook)
	5	(a) 70.82; 76.75; 16.89; (b) 708.2, 767.5, 168.9, 7.082, 7.675, 1.689; 7082, 7675, 1689
	6	2.88389; 1.94052; $\overline{3}$.94151
	7	792.4; 17.49; 1.514; 486.5
	8	4.87614; 1.62363
	9	67.603; 4.7547
	10	146.17; 36.8
5	11	9.88; 5.422; 5.208
	12	0.2783
	13	0.0000001432
	14	237.6
	15	187.08
	16	14.403 square inches
	17	2.203 or, say, $2\frac{1}{4}$ inches
	18	107 horsepower
	19	Page **121**
	20	No
	1	8001.3 cubic inches
	2	83.905 square inches
	3	69.395 cubic inches
	4	1.299 inches
	5	22.516 cubic inches
	6	8 inches
	7	0.0276 cubic inches
6	8	4.2358 inches
	9	1.9635 cubic inches
	10	410.5024 cubic inches
	11	26.4501 square inches
	12	Radius 1.4142 inches: area, 0.43 square inch
	13	Area, 19.869 square feet; volume, 10.2102 cubic feet
	14	Area, 240 square feet; volume, 277.12 cubic feet

Section Number	Number of Question	Answers (Or where information is given in Handbook)
6	15	11.3137 inches
	16	41.03 gallons
	17	17.872 square inches
	18	1.032 inches
	19	40 cubic inches
	20	Table page 169
	21	Table page 169
	22	5.0801 inches
	23	4 inches; 5226 inches
7	1	Page 277
	2	Page 277
	3	Page 277
	4	Page 277
	5	Page 278
	6	Page 278
	7	Page 278
	8	Page 278
	9	Page 279
	10	Page 279
	11	Page 279
	12	Page 279
	13	Page 281
	14	Page 283
	15	Page 280
	16	Page 280
	17	Page 280
	18	Page 280
	19	Page 280
	20	Page 281
	21	Page 281
	22	Page 281
	23	Page 282
	24	Page 282
	25	Page 282
	26	Page 282

Section Number	Number of Question	Answers (Or where information is given in Handbook)
8	1	See page 178
	2	In any right-angle triangle having an acute angle of 30 degrees, the side opposite that angle equals 0.5 × hypotenuse.
	3	Sine = 0.31634; tangent = 0.51549; cosine = 0.83942
	4	Angles equivalent to tangents are 27° 29′ 24″ and 7° 25′ 16″; angles equivalent to cosines are 86° 5′ 8″ and 48° 26′ 52″
	5	Rule 1: Side opposite = hypotenuse × sine, Rule 2: Side opposite = side adjacent × tangent
	6	Rule 1: Side adjacent = hypotenuse × cosine, Rule 2: Side adjacent = side opposite × cotangent
	7	Page 171
	8	Page 172
	9	After dividing the isosceles triangle into two right-angle triangles
	10	Page 171
9	1	2 degrees 58 minutes
	2	1 degree 47 minutes
	3	2.296 inches as shown by the table on page 77
	4	$\dfrac{360°}{N} - 2a$ = angle intercepted by width W. The sine of $\frac{1}{2}$ this angle × $\frac{1}{2} B$ = $\frac{1}{2} W$; hence, this sine × $B = W$
	5	3.1247 inches
	6	3.5085 inches
	7	1.7677 inches
	8	75 feet approximately

Section Number	Number of Question	Answers (Or where information is given in Handbook)
9	9	a = 1.0316 inch; b = 3.5540 inches; c = 2.2845 inches; d = 2.7225 inches
	10	a = 18° 22'. For solution of similar problem, see Example 4 of Section 9
	11	A = 5.8758"; B = 6.0352"; C = 6.2851"; D = 6.4378"; E = 6.1549"; F = 5.8127" Apply formula on Handbook page 176
	12	2° 37' 33"; 5° 15' 6"
	13	5.2805 inches
	14	10 degrees 23 minutes
10	1	84°; 63° 31'; 32° 29'
	2	B = 29°; b = 3.222 feet; c = 6.355 feet; area = 10.013 square feet
	3	C = 22°; b = 2.33 inches; c = 1.358 inches; area = 1.396 square inches
	4	A = 120° 10'; a = 0.445 foot; c = 0.211 foot; area = 0.027 square feet
	5	The area of a triangle equals one-half the product of two of its sides multiplied by the sine of the angle between them. The area of a triangle may also be found by taking one-half of the product of the base and the altitude
11	1	Page 267; page 262
	2	Page 252; page 266
	3	Page 266; page 232
	4	30° 30'; 45° 5'
	5	28° 21' 30"; 22° 25' 40"
	6	5.2
	7	81.16 pounds
	8	16.5604

Section Number	Number of Question	Answers (Or where information is given in Handbook)
12	1	Page 1679 for Morse
		Page 1683 for Jarno
		Page 1691 for milling machine
		Page 1137 for taper pins
	2	2.205 inches; 12.694 inches
	3	4.815 inches. Page 1560
	4	1.289 inches. Page 1561
	5	3.110 inches. Page 1561
	6	0.0187 inch
	7	0.2796 inch
	8	1.000 inch
	9	26 degrees 7 minutes
13	1	Pages 1509, 1511
	2	Page 1507
	3	Page 1507
	4	Page 1506
	5	Page 469
	6	Page 470
	7	Page 1508
	8	Page 1257
	9	Pages 1255, 1257
	10	When the tolerance is unilateral
	11	See page 1518
	12	It means that a tolerance of 0.0004 to 0.0012 inch could normally be worked to. See table of American Standard Tolerances on page 1517
14	1	(a) and (b) Both countries now use the Unified Standard. See pages 1255 and 1338
	2	This is the symbol that is used to specify an American Standard screw thread 3 inches in diameter, 4 threads per inch or the coarse series, and Class 2 fit.
	3	Acme thread is stronger, easier to cut

Section Number	Number of Question	Answers (Or where information is given in Handbook)
14		with a die, and more readily engaged by a split nut in the case of lead-screws.
	4	The Stub Acme form of thead is preferred for those applications where a coarse thread of shallow depth is required
	5	See tables, pages 1262, 1263
	6	$\frac{3}{4}$ inch per foot measured on the diameter—American and British standards
	7	Page 1369
	8	Center line of tool is set square to axis of screw thread
	9	Present practice is to set center line of tool square to axis of pipe
	10	See formulas for F_{rn} and F_{rs}, page 1319
	11	By three-wire method or by use of special micrometers; see pages 1373 to 1389
	12	Two quantities connected by a multiplication sign are the same as if enclosed by parentheses; see instructions about order of operations, page 95
	13	(a) Lead of double thread equals twice the pitch; (b) lead of triple thread equals three times the pitch; see page 1373
	14	See pages 1257 and 1259
	15	0.8337 inch; see page 1379
	16	No. Bulk of production is made to American Standard, dimensions given in Handbook
	17	This standard has been superseded by the American Standard
	18	Most machine screws (about 80% of the production) have the coarse series of pitches

Section Number	Number of Question	Answers (Or where information is given in Handbook)
	19	(a) Length includes head; (b) length does not include head
	20	No. 25; see table, page 1406
	21	0.1935 inch; see table, page 1616
	22	Yes. The diameters decrease as the numbers increase
	23	The numbered sizes range in diameter from 0.0059 to 0.228 inch, and the letter sizes from 0.234 to 0.413 inch; see pages 1615 to 1617
	24	A thread of ¾ standard depth has sufficient strength, and tap breakage is reduced
14	25	(a) and (b) the American (National) form — formerly U. S. Standard
	26	Cap-screws are made in the same pitches as the Coarse-, Fine-, and 8-thread series of the American Standard, Class 2A
	27	American National form (formerly United States Standard) four standard diameters; pages 1362 and 1363
	28	Page 1640
	29	Page 1640
	30	0.90 × pitch; see pages 1376, 1377
	31	See last paragraph, page 1674
	32	Page 1674
	33	Included angle is 82° in each case
15	1	A foot-pound in mechanics is a unit of work and is the work equivalent to raising one pound 1 foot high
	2	1000 foot-pounds
	3	Only as an average value; see page 331
	4	28 foot-pounds; see pages 328 and 329

Section Number	Number of Question	Answers (Or where information is given in Handbook)
15	5	1346 pounds
	6	Neglecting air resistance, the muzzle velocity is the same as the velocity with which the projectile strikes the ground; see page 324
	7	See page 293
	8	Square
	9	1843 pounds approximately
	10	The pull would have been increased from 1843 pounds to about 2617 pounds; see page 305
	11	Yes
	12	About 11 degrees
	13	The angle of repose
	14	The coefficient of friction equals the tangent of the angle of repose
	15	32.16 feet per second per second
	16	No. 32.16 feet per second is the value at sea level at a latitude of about 40 degrees, but this figure is commonly used — see page 290
	17	No. The rim stress is independent of the diameter and depends upon the velocity; see page 344
	18	The disruptive force = centrifugal ÷ 3.1416
	19	No. The increase in stress is proportional to the square of the rim velocity
	20	110 feet per second or approximately 1 mile per minute
	21	Because the strength of wood is greater in proportion to its weight than cast iron
	22	See page 351
	23	In radians per second

Section Number	Number of Question	Answers (Or where information is given in Handbook)
15	24	A radian equals the angle subtended by the arc of a circle when the length of the arc equals the radius of the circle; this angle is 57.3 degrees nearly
	25	Page 276
	26	60 degrees; 72 degrees; 360 degrees
	27	Page 275
	28	Page 354 (see page 353 for example illustrating method of using table)
	29	Length of arc = radians × radius. As radius = 1 in table of segments, l = radians
	30	40 degrees, 37.5 minutes
	31	176 radians per second; 1680.7 revolutions per minute
	32	1.5705 inches
	33	27.225 inches
16	1	Page 358
	2	12,000 pounds
	3	1 inch
	4	Page 446
	5	Page 357
	6	Page 446
	7	440 pounds
	8	3-inch diameter; see page 431
17	1	1.568 (see formula on Handbook page 453)
	2	6200 pounds per square inch approximately
	3	It depends upon the class of service; see page 460
	4	Tangential load = 550 pounds; twisting moment, 4400 inch-pounds

Section Number	Number of Question	Answers (Or where information is given in Handbook)
17	5	See formulas on page 455 and also the table on page 455
	6	The head is useful for withdrawing the key, especially when it is not possible to drive against the inner end; see page 978
	7	Key is segment-shaped and fits into circular keyseat; see pages 984, 985
	8	These keys are inexpensive to make from round bar stock, and keyseats are easily formed by milling
	9	0.211 inch; see table, page 983
18	1	18 teeth; 3 inches; 0.2618 inch
	2	2.666 inches; 2.333 inches; 0.1666 inch
	3	Pages 739 and 740
	4	Chordal thickness at intersections of pitch circle with sides of tooth
	5	Table, pages 737 and 738
	6	Table, page 744
	7	Surface durability stress and tooth fillet tensile stress are the two principal factors to be found in determining the power transmitting capacity of spur gears. Formulas and procedures will be found on pages 825 through 830
	8	Because the tooth shape varies as the number of teeth is changed
	9	No; one hob may be used for all tooth numbers, and the same applies to any generating process
	10	Stub
	11	Page 742 under Fellows Stub Tooth

Section Number	Number of Question	Answers (Or where information is given in Handbook)
	12	Page 951
	13	Page 759
	14	Page 745
	15	See table on page 741
	16	Pages 1417, 1420 to 1443
	17	Pages 777 to 780
	18	Page 952
	19	Pages 852, 855, 857
	20	Yes, but accurate tooth form is obtained only by a generating process
	21	See paragraph on page 880
	22	14 teeth; see page 852
	23	Yes, 13. See pages 852 and 856 and compare the ratios and tooth numbers
	24	Page 851
	25	See text on face angles on page 852
18	26	When the number of teeth in both the pinion and the gear are the same, the pitch angle being 45 degrees for each
	27	The whole depth minus the clearance between the bottom of a tooth space and the end of a mating tooth = the working depth
	28	See page 747
	29	Pages 823 and 824
	30	See pages 734 and 737
	31	See diagram, page 851
	32	Circular pitch of gear equals linear pitch of worm
	33	Pitch diameter; see Rule number 12, page 896
	34	Helix angle or *lead* angle of worm is measured from a plane perpendicular to the axis; helix angle of a helical gear is measured from the axis

Section Number	Number of Question	Answers (Or where information is given in Handbook)
18	35	These terms have the same meaning in each case
	36	Table, pages 908 and 909
	37	$3\frac{1}{4}$ inches and $1\frac{1}{4}$ inches; use table, pages 908 and 909 in reverse order
	38	Page 902
	39	To provide clearance space between worm and gear and a grinding allowance, see page 913
	40	Page 913
	41	See explanation beginning page 895
	42	Page 898
	43	Page 917
	44	Normal diametral pitch is commonly used
	45	Yes (see page 917), but the hobbing process is generally applied
	46	Pitch diameter
	47	Page 943
19	1	The number indicates the kind of steel and approximately its nickel and carbon content; see pages 2014 and 2021
	2	The diameter of each end and the length of taper; see explanation on page 1678; also table, 1683
	3	Tolerance is applied in whatever direction is likely to be the least harmful; see page 1507
	4	It is said that James Watt found, by experiment, that an average cart-horse can develop 22,000 foot-pounds per minute, and added 50 per cent to insure good measure to purchasers of his engines (22,000 × 1.50 = 33,000)

Section Number	Number of Question	Answers (Or where information is given in Handbook)
19	5	Tin in the higher grades, and lead in the lower grades
	6	Same depth as an ordinary gear of 10 diametral pitch
	7	The tooth thickness and the number of teeth is the same as an ordinary gear of 8 diametral pitch
	8	Add 2 to the number of teeth and divide by the outside diameter
	9	Multiply the outside diameter by 3.1416 and divide the product by the number of teeth, plus 2
	10	Birmingham or Stub's iron wire gage is used for seamless steel, brass, copper, and aluminum tubing
	11	Iron wire rope has the least strength of all wire rope materials. See page 477
	12	If surfaces are well lubricated, the friction is almost independent of the pressure, but if the surfaces are unlubricated, the friction is directly proportional to the normal pressure excepting for the higher pressures
	13	It depends very largely upon temperature. See "lubricated surfaces," page 541
	14	$8 - (- 4) = 12$. See rules for positive and negative numbers, page 105
	15	Yes. One meter equals 3.2808 feet; see other equivalents on page 2361
	16	Experiments have shown that there is definite relationship between heat and work and that one British thermal unit equals 778 foot-pounds. To change 1 pound of water at 212 deg. F. into steam at that temperature re-

Section Number	Number of Question	Answers (Or where information is given in Handbook)
19		quires about 966 British thermal units or $966 \times 778 = 751,600$ foot-pounds nearly; hence the number of pounds of water evaporated at 212 deg. F. equivalent to 1 horsepower-hour $= 1,980,000 \div 751,600 = 2.64$ pounds of water as given in the Handbook, page 2379
	17	No. The thickness of the pipe wall is increased by reducing the inside diameter; compare thickness in table on page 2246
	18	Wrought iron was preferred at one time, but tests have shown that for general work it is equaled by steel pipe. See page 2249
	19	Yes. The so-called "O.D. pipe" begins, usually, with the 14-inch size
	20	A fitting having a 45-degree branch; see illustrations on pages 2256 and 2257 of a lateral and a reducing lateral
	21	Yes. About 140 degrees lower; see page 2226
	22	Low-carbon alloy steels of high-chromium content; see page 2035
	23	A low-carbon steel containing 0.20% carbon or less and usually from 0.60 to 0.90% manganese; see page 2025
	24	No. The nominal length of a file indicates the distance from the point to the "heel" and does not include the tang
	25	Yes. See table page 2193
	26	Yes. Certain compositions of bismuth, lead, tin, and cadmium; see page 2194

Section Number	Number of Question	Answers (Or where information is given in Handbook)
19	27	(a) and (b) A number indicating how a given volume of the material or liquid compares in weight with an equal volume of water. (c) A number indicating a comparison in weight with an equal volume of air; see pages 2195 and 2196
	28	The first digit identifies the alloy type; the second, the impurity control; etc. See page 2166
	29	Red brass contains 84 to 86% copper, about 5% tin, 5% lead, and 5% zinc, whereas yellow brass contains 62 to 67% copper, about 30% zinc, 1.5 to 3.5% lead and not over 1% tin; see S.A.E. Specifications 40 and 41 on page 2155
	30	See pages 2232 and 2233
	31	No. Twenty-six sizes ranging from 0.234 to 0.413 inch are indicated by capital letters of the alphabet (see table, pages 1616, 1617). Fractional sizes are also listed in manufacturers' catalogues beginning either at $\frac{1}{32}$-inch, $\frac{1}{16}$-inch, or $\frac{1}{8}$-inch, the smallest size varying with different firms
	32	To insure uniform heating at a given temperature and protect the steel against oxidization; see page 2052
	33	Hardening temperatures vary for different steels; see critical temperatures and how they are determined, pages 2050 and 2051
	34	Set the taper attachment to an angle, the cosine of which equals 0.125 ÷ 0.1255; see page 1456

Section Number	Number of Question	Answers (Or where information is given in Handbook)
19	35	See page 1545
	36	Divide $\frac{3}{4}$ by 12; multiply the taper per inch thus found by 5 and subtract the result from the large diameter; see rules for figuring tapers, page 1557
	37	Metric horsepower is 450 foot-pounds less; see page 289
	38	0.88333; see page 277
	39	$x = 3$
	40	About $12\frac{1}{2}$ degrees; see page 701
	41	Ratio between resistance to the motion of a body due to friction, and the perpendicular pressure between the sliding and fixed surfaces; see formula, page 541
	42	No. Stub's steel wire gage applies to tool steel rod and wire, and the most important applications of Stub's iron wire gage (also known as Birmingham) are to seamless tubing, steel strips and telephone and telegraph wire
	43	If the difference between the lengths of the pawls equals one-half of the pitch of the ratchet wheel teeth, the practical effect is that of reducing the pitch one-half; see ratchet gearing, page 975
	44	Apply belt with hair or "grain" side next to pulleys and driving side below so that the slack side above increases arc of contact and pulling power
	45	For a quarter-turn drive, align the center of the driven pulley face with that face of the driving pulley from which the belt leaves; see Angular Drive, page 1042

Section Number	Number of Question	Answers (Or where information is given in Handbook)
19	46	The ultimate strength is less due to bending action; see formula, page 1113 and also table 1115, "Approximate Breaking Strain."
	47	The studded chain has less strength than an unstudded chain; page 1114
	48	Multiply 90 by 12 and divide by the circumference of the shaft to obtain R.P.M.; see cutting speed calculations, pages 1703 and 1704
	49	(a) Lard oil; (b) gasoline
	50	If the teeth advance around the gear to the right, as viewed from one end, the gear is right-handed; and, if they advance to the left, it is a left-hand gear; see illustrations, page 918
	51	No. They may be of opposite hand depending upon the helix angle; see pages 919 and 920
	52	Multiply the total length by the weight per foot for threaded and coupled pipe, given in table, page 2246
	53	The processes are similar but the term "pack-hardening" usually is applied to the casehardening of tool steel; see pages 2062 and 2063
	54	A gas process of surface hardening; see page 2064
	55	See definitions for these terms given on page 1876
	56	About 34 inches but the height may vary from 32 to 36 inches for heavy and light assembling, respectively
	57	Major diameter is the same as outside diameter and minor diameter is the same as root diameter; see definitions, page 1256

Section Number	Number of Question	Answers (Or where information is given in Handbook)
19	58	The S.A.E. Standard conforms, in general, with the Unified and American Standard Screw Thread Series as revised in 1959 and may, therefore, be considered to be the same for all purposes
	59	See information on work materials, pages 1697 and 1698
	60	Yes. See pages 442 and 2043
	61	By the open-hearth process. For general description of this process see page 1983
	62	The finer grades of tool steel are made by the crucible and electric processes, but most tool steel used for such tools as hammers, pliers, picks, axes, knives, agricultural implements, etc., is made by the open-hearth process
	63	According to the table on page 1706, the recommended cutting speed is 300 feet per minute. Since this is for average conditions and is intended as a starting point, it is important to know the factors which affect cutting speed as covered in the introductory section on page 1697.
	64	No. First determine the diametral pitch the same as for a spur gear; then divide this "real diametral pitch" by the cosine of the helix angle to obtain the "normal diametral pitch" which is the pitch of the cutter; see page 917
	65	The Fellows system is based upon diametral pitch and the Nuttall system upon circular pitch

Section Number	Number of Question	Answers (Or where information is given in Handbook)
	66	In making gear patterns for cast gears and when fairly accurate profiles must be drawn
	67	The face and flank of a tooth consists of circular arcs, the radii of which are obtained from the odontograph table; see table, page 824, and instructions on page 823
	68	No, The size of the gear blank, the pitch of the teeth and depth of cut are sufficient for the man in the shop. The tooth curvature is the result of the gear-cutting process. Tooth curves on the working drawing are of no practical value
	69	By changing the inclination of the dividing head spindle
19	70	See formula and example on page 717
	71	Divide the total number of teeth in both gears by twice the diametral pitch to obtain the theoretical center-to-center distance. (See formula in the table of Formulas for Dimensions of Standard Spur Gears, page 746)
	72	Subtract number of teeth on pinion from number of teeth on gear and divide the (See Rule at bottom of page 807)
	73	See page 1240
	74	The Standard Wire Gage (S.W.G.), also known as the Imperial Wire Gage and as the English Legal Standard, is used in England for all wires
	75	A simple type of apparatus for measuring power

Section Number	Number of Question	Answers (Or where information is given in Handbook)
	76	With a dynamometer, the actual amount of power delivered may be determined; that is, the power input minus losses. See page 710
	77	The uniformly loaded beam has double the load capacity of a beam loaded at the center only; see formulas, page 418
	78	"Dense" southern yellow pine is stronger; see table, page 447
	79	No. The nominal size of steel pipe excepting sizes above 12 inches, is approximately equal to the inside diameter; see tables, pages 2246 and 2247
	80	0.357 inch. See formula, page 382
	81	Because log 10 = 1, and log 100 = 2; hence, logs of all intervening numbers equal 1 plus some fraction
19	82	(a) 1.04139; (b) 1.99564
	83	Multiply depth in feet by 0.4335; see table of constants, weight in pounds of water column, page 81
	84	(a) 760 pounds, page 1744. (b) 0.42 hp by interpolation, page 1744
	85	100 per cent
	86	50 per cent
	87	Various steels are used depending on kind of spring. See page 2032
	88	This is a special annealing process. The steel is heated above the critical range and allowed to cool in still air at ordinary temperature, page 2061. Normalizing temperatures for steels are given on pages 2071 and 2073
	89	The standard mounting dimensions, frame sizes, horsepower and speed ratings. See section beginning on page 2295

Section Number	Number of Question	Answers (Or where information is given in Handbook)
19	90	Yes. The American standard drafting-room practice includes section lining, etc. See page 2333
	91	No. There are different tapers per foot ranging from 0.5986 to 0.6315 inch; see table, page 1679
	92	Yes. See page 1410
	93	Unilateral and plus; see page 1507
	94	See table, page 1324
	95	If D = diameter of hole in inches; T = stock thickness; shearing strength of steel = 51,000 pounds per square inch, then tonnage for punching = $\dfrac{51,000\ D\pi T}{2000}$ = 80 DT
	96	See page 465
	97	The Brown & Sharpe or American wire gage is used in each case; see pages 463 to 465
	98	No, this name is applied to several compositions which vary widely
	99	Antimony and copper
	100	177 nearly; see table beginning on page 1791
	101	See page 2193
	102	See pages 1769 to 1771
	103	See page 1755
	104	See page 1852
	105	See page 1456
	106	See pages 1595 and 1607
	107	The practice varies widely and includes low-carbon and high-carbon steels; also special alloy steels; see page 1831
	108	The terms "go" and "not go" are preferable for limit gages; see page 1849

Section Number	Number of Question	Answers (Or where information is given in Handbook)
19	109	The lead of a milling machine equals lead of helix or spiral milled when gears of equal size are placed on feed screw and worm gear stud; see rule for finding lead on page 1457
	110	Multiply product of driven gears by lead of machine and divide by product of driving gears. If lead of machine is 10, divide 10 times product of driven gears by product of drivers
	111	$\frac{5}{11}$; $\frac{79}{188}$; $\frac{18}{25}$
	112	The whole depth and tooth thickness at the large ends of the bevel gear teeth are the same as the whole depth and thickness of spur gear teeth of the same pitch
	113	See text on page 1878
	114	5.7075 inches
	115	Use the formula (Handbook page 1113) for finding the breaking load which in this case is taken as three times the actual load. Transposing, $D = \sqrt{\dfrac{6 \times 2000 \times 3}{54,000}} = 0.816$ inch, say $\frac{7}{8}$ inch diameter
	116	Because the direction of the cutter thrust tends to cause the gear to rotate upon the arbor; see page 938, "Milling the Helical Teeth"
	117	Due to its collapsing action, the tap may be removed rapidly from the hole immediately after the tapping is completed; see page 1667
	118	Chiefly when a hole either is to be tapped or reamed after drilling; see page 1883

Section Number	Number of Question	Answers (Or where information is given in Handbook)
	119	See table on page 1998
	120	Page 1094
	121	See "Control of Surface Roughness" on page 2334
	122	See pages 838, 874, and 875
	123	See page 1654
19	124	See table on page 1943
	125	See table on page 1938 and also text on pages 1938 and 1939
	126	Motor rotation has been standardized by the National Electrical Manufacturers Association. See page 2296
	127	Yes. See last paragraph on page 688

INDEX

PAGE